SURVIVING THE APOCALYPSE

Understanding and fighting through the coming emergency

Canada

Guernica Editions Inc. acknowledges the support of the Canada Council for the Arts and the Ontario Arts Council. The Ontario Arts Council is an agency of the Government of Ontario.

We acknowledge the financial support of the Government of Canada.

SURVIVING THE APOCALYPSE

Understanding and fighting through the coming emergency

Thomas F. Pawlick

MiroLand
publishers

MIROLAND (GUERNICA)
TORONTO • CHICAGO • BUFFALO • LANCASTER (U.K.)
2021

Connie McParland, series editor
Michael Mirolla, editor
David Moratto, cover and interior design
Tamara Bauer, Front cover image
Guernica Editions Inc.
287 Templemead Drive, Hamilton, ON L8W 2W4
2250 Military Road, Tonawanda, N.Y. 14150-6000 U.S.A.
www.guernicaeditions.com

Distributors:
Independent Publishers Group (IPG)
600 North Pulaski Road, Chicago IL 60624
University of Toronto Press Distribution,
5201 Dufferin Street, Toronto (ON), Canada M3H 5T8
Gazelle Book Services, White Cross Mills
High Town, Lancaster LA1 4XS U.K.

First edition.
Printed in Canada.

Legal Deposit—First Quarter
Library of Congress Catalog Card Number: 2020934842
Library and Archives Canada Cataloguing in Publication
Title: Surviving the apocalypse : understanding and fighting through the coming emergency / Thomas F. Pawlick.
Names: Pawlick, Thomas, author.
Description: First edition.
Identifiers: Canadiana (print) 20200208985 | Canadiana (ebook) 20200208993 | ISBN 9781771835749 (softcover) | ISBN 9781771835756 (EPUB) | ISBN 9781771835763 (Kindle)
Subjects: LCSH: Emergency management. | LCSH: Emergency management—Citizen participation. | LCSH: Hazard mitigation. | LCSH: Hazard mitigation—Citizen participation. | LCSH: Climatic changes. | LCSH: Social ecology. | LCSH: Human ecology. | LCSH: Social action. | LCGFT: Essays.
Classification: LCC HV551.2 .P39 2021 | DDC 363.34/6—dc23

Dedicated to a little girl named Hope.

Sensing subversion

I want to make you uneasy,
suspicious.

sensing subversion
somewhere, somehow
out of step
out of line
deviating from the straight line

hunt it down

to see you
curl your lip. Sneer.
powerful
righteous
Born Again, holier-than-thou
now
knowing it all,
tall in the saddle
rock-ribbed, reciting

then slap you,
feel the back of my hand
on safety
and slogans,
knock you
right on over

I want to wake you
up.

CONTENTS

Introduction

YOU COULD DIE

It's coming at us like a freight train. Not since the first small group of *homo sapiens*, somewhere in Africa 200,000 years ago, struggled for survival against the odds, have we, as a species, been in such danger.

How many were there, then? A couple of dozen? The very first ones, Adam and Eve and the hominid kids. Almost anything could have wiped them out, in Africa, with its super potent bugs: an outbreak of dengue fever, or malaria, or blackwater fever, or an infected tooth. Or death from drought, slow starvation on the parched veldt. A persistent big carnivore, or a pack of proto-hyenas might have finished them quickly, in a matter of days or even hours. End of story.

But they made it. Had strong antibodies, or found water, or reacted fast, jumped out of the way of tooth and claw, and climbed a tree. Alert. Maybe they had help from a Creator, maybe not. But at least they did their own part.

Not like us, sitting here on the railroad tracks all these centuries later, with our lazy backs turned, munching a double bacon cheeseburger and watching the Super Bowl on TV while destruction bears down. Oblivious, like the Romans of Juvenal's day, "anxious for just two things: bread and circuses."

Oblivious to what?

To ourselves, to Newton's Third Law of Motion, and to the coming storm our very success has provoked. In other words, failing to understand and anticipate the idea of resistance.

Systems and Resistance

Newton's third law states that "for every action there is an equal and opposite reaction." Hit a racquetball hard against the wall and, except for the minuscule drag of friction with the air, it will rebound back just as hard. The wall has offered resistance.

Sir Isaac was only talking about physical things, of course, but the modern exponents of Systems Theory have gone him one better, showing that his law turns out to apply almost everywhere, including in the complex ecological workings of the natural world and in human social, political and economic relations, where one may observe all sorts of resistance and reaction.

Systems Theory, first proposed by German biologist Ludwig von Bertalanffy and later expanded by ecologists, anthropologists and now some mainstream economists, states that everything that exists—from the smallest sub-atomic particles to the largest galaxies—is organized in individual, yet interlocked systems, which themselves are part of still bigger systems.

Even chaos, if certain physicists are right, apparently has its secret system.

It's hard not to agree with the empirical basics of the thing. The evidence plays itself out before our eyes every day. In school, we all had to study math, which is nothing if not systematic, and most of us took high school chemistry, in which was explained the decreasing hierarchies of matter and energy, of molecule, atom, sub atomic particles, energy waves, and so forth. Experiments in the chem lab showed us how the latter were connected, and end-of-term math exams made it all too plain how algebra, geometry, trig and calculus all supposedly meshed together, though unfortunately not always in our brains, which are also systems, though not necessarily functioning ones when exams require it.

Other systems are equally obvious. There are social and political systems, which run from the basic family, to tribes and clans, to villages, to cities, countries and so on. Municipal councils (supposedly) mesh with county ones, and those in turn with the larger systems of state or provincial legislatures, which presumably harmonize with national legislatures, and eventually with the undisciplined mishmash of international treaty regimes. Whether the meshing and harmonizing is top-down or bottom-up in nature varies.

Every thinker from Plato and Aristotle on down has been able to see a diversity of systems. But what formal Systems Theory does is make the subsystems and their relations to each other much clearer, much more obvious. It gives the way they work together *rules*. Rules that appear logically inviolate, in fact—almost—inexorable. One of those rules, borrowed from Systems Theory's first cousin, Chaos Theory, is that even very small changes in some systems can have very big effects in others (the so-called Butterfly Effect), while changes in many systems at once, each impacting the others and being impacted by them, can provoke a wave of disturbance: a Cascade Effect. These effects result from each system's internal positive and negative "feedback loops," the positive ones increasing a system's output, and the negative ones decreasing it.

Such rules, applied to right here, right now, on planet Earth, should scare hell out of you, because some of the changes people have been making to the world around us are about to provoke truly massive resistance reactions—massively destructive ones. So destructive, that if they *don't* scare you, there is something wrong with your mind. And their simultaneous arrival is far closer than most scientists, looking only at their individual specialties, have foreseen.

Here, you may yawn, and think (pick one): "Oh, another pointy-headed lefty, predicting doom." Or: "Oh, another born-again righty, predicting the End Times." A nut.

If so, you have not been paying attention to the growing presence of resistance in our world. It is like a time bomb, hidden in a tangled thicket, the potent fruit of a vine that twines and winds its way through story after story, first on one opposing side, then the other, as if acting out some subtle parable of nature. It will, eventually, be our doom or our salvation, particularly that of the so-called "millennials," who came of age with the turn of the century, and anyone born after the year 2000. They will in their lifetimes be faced with the most fearsome crisis in history.

The process of self destruction

It started as a set of man-made systems, developed over generations, with what once were useful purposes:

a) to promote or maintain physical health (medical system);
b) to produce nourishing food (agricultural system);
c) to rationally organize social action and exchange (economic system);
d) to provide heat for physical comfort and power to perform work (energy system).

Over the centuries, however, the original aims of those systems were perverted, at first subtly then rapidly, giving way to a single, overriding goal: to amass money. The process accelerated particularly fast in the past 100 years, while the directional control at the top of each system also changed, from being operated by individuals or by local communities, to being controlled by ever larger and larger, and ever fewer corporate entities.

These, in turn, pressed harder and harder to amass yet more money, forcing the systems to work faster through a myriad positive feedback loops, and jettisoning any negative loops that might slow the process. That process, spurred by what has been dubbed the "Magic Bullet" mentality, has become increasingly strong—even violent—provoking increasingly strong resistance.

Raquetballs have started rebounding off walls. What once was beneficial has become outright harmful, racing toward the opposite of the systems' original goals by:

a) weakening or destroying physical health, and killing people;
b) destroying the capacity to produce nourishing food, and ruining the soil;
c) producing social and economic ruin, particularly of the young;
d) threatening to destroy the atmosphere and extinguish our planet's life via climate change and radiation from nuclear waste.

There are, of course, many other, similar examples, as many as there are human systems. This process is going on across the board, reacting to the same causes and following the same patterns. But to try to describe the historic degradation of every system, great or small, would require an encyclopaedia-length work. This book will look only at the ongoing collapse of the four fundamental systems on which our lives, and the lives of our children directly depend: the medical, agricultural, economic and energy systems.

And at the eventual, cataclysmic cascade of their effects on each other in the overall planetary system that is our biosphere. The dam is closer to bursting than anyone had anticipated.

Let's start with the resistance of certain harmful bacteria—the kind that cripple and kill people—to antibiotics. Right now, today, if you are hurt or fall ill and are rushed to any emergency ward in any hospital in North America, you run a serious risk of encountering one or another of these resistant microbes, being infected by it, and of its not going away when treated with standard medications.

You could die.

Part I
THE CRUCIAL SYSTEMS

Chapter One

Zombie bugs

Ring around the rosie, pocket full of posies,
ashes, ashes, we all fall down.[1]

Most of medical history has focused on three things: 1) fixing broken body parts, 2) preventing the broken stuff from getting infected, or, as they used to say in the old, old days, "putrifying," and 3) fighting "fevers," which often came associated with assorted rashes, bumps and festering pustules. As medicine started to get better organized and, thanks to the invention of the microscope, the causes of putrefaction and fevers became clearer, doctors started referring to numbers two and three as "infection control."

They still do, although these days it's more like infection out-of-control, with the advent of a ballooning population of antibiotic-resistant bacteria, and a simultaneous abdication, especially in North America, of business and government efforts to deal responsibly with the crisis.

In fact, one could say our entire medical care system has itself been infected, not with microbes, but with a mindset that risks destroying our chief means of physical defense against bacterial disease, while blocking an existing, thoroughly tested alternative solution that could supplement or replace antibiotics. This attitude also threatens the research structure from which come most new treatments and drugs, and compromises the effectiveness of regulatory bodies charged with assuring the safety of both drugs and medical procedures. It prices drugs out of the reach of patients, in some cases consigning whole groups to a choice between bankruptcy or death.

As will be seen in chapters three, five and six, it also erodes the altruistic character of the educational system that trains doctors, nurses, researchers, medical technicians and hospital administrators. It drives graduates into a form of debt servitude that curbs their scientific and social independence. It forces inappropriate and coercive treatments on some patients, including the mentally ill, and helps to impoverish society by encouraging business tax dodges on a massive scale.

Several traits mark this mindset, including a persistent belief in "magic bullet" quick fixes, and a kind of tunnel vision that fails to consider the long term, knock-on effects of its actions. It is impatient, heavy-handed, and gravitates toward simplistic, all-or-nothing approaches to problems. But its chief characteristic is a single-minded focus on monetary profit as the overriding motive for all medical decisions. The latter has roots in the earliest years of medical history, but has bloomed like a malevolent weed with the coming to dominance of today's multinational corporations, and the concentration in their hands of the economic and political control of society.

By its nature, it both exhibits and provokes resistance.

Factor in the myriad ways in which our medical structures affect and are affected by other basic systems, from food production to climate, and the potential for mischief is vast.

To understand the processes that brought this about, and identify ways to reverse them, requires a look—using the clarifying lens of Systems Theory—at the past, at the original goals of medicine and medical education, and how the seeds of subversion were planted.

Crude and ineffective

Early efforts to prevent local infections were sometimes crude and often ineffective. Ancient Rome's doctors passed surgical instruments through an open flame before using them to operate, a primitive form of disinfection, but it's uncertain what led them to do it, or what, exactly, they thought it accomplished. When they had to amputate a Centurion's arm or leg after a battle, they also cauterized the stump with hot irons. This did have the positive effect of stopping the bleeding, by shrinking and closing off blood vessels. But once the sizzling stopped, the burnt, dead tissue around the wound actually became a prime site for bacterial invasion, increasing rather than decreasing the likelihood of putrefaction. Most soldiers who weren't killed by their wounds in battle died later of infection. That is, if they didn't die even before battle of fever or chills contracted after marching too long in the rain and living in unsanitary conditions in the field.

It was also vaguely understood, in China since at least 1117, that boiling food or wine might prevent their putrefaction during storage, but no one was sure why. And of course one couldn't boil people to save them from infection, although some Chinese emperors may have boiled them for other reasons.

By the time of the arrival of the international disaster that became known as the Black Death, in 1348, control techniques hadn't advanced much. Bubonic plague, an infection by the bacteria *Yersinia pestis*, whose symptoms included a blackish rash, or "rosie," followed by the appearance of egg-like "buboes" in the armpit, groin or neck, debuted by killing well over one third of Europe's population, including 30 million peasants, in only two years. Spread by fleas that lived on the black rat (*Rattus rattus*), it went on killing after that in repeating cycles, wiping out millions more, until it finally subsided in the late 1600s.

Some ecological investigators, though agreeing that fleas were the immediate disease vector, dispute the black rat's key role. There is evidence that climate change, which caused serious drought in Central Asia, led to an actual drop in rodent numbers—specifically in populations of marmots and ground squirrels native to China, which host the flea.[2] This forced the fleas to seek other warm blooded hosts, including the humans and camels in trading caravans travelling the Silk Road. European traders may thus have brought the fleas home with them, introducing them to other hosts, like the rat. The possibility that climate change in Asia may have been at the ultimate root of the whole thing has particularly ominous implications today, as later chapters will show.

Except for the idea of quarantine, introduced in Venice where incoming ships with sick aboard were held offshore for 40 days (*quaranti giorni* in Italian) before any passengers could land, most conscious efforts at prevention or control of plague were useless. This was because no one really had any idea how infections spread. The prevailing theory at the time was that it was caused by "miasma," or "bad air," produced by the rotting of dead things. Thus professional perfumers and fumigators, dressed in white robes with red crosses (precursors of the Red Cross), roamed about, "smoking" the bad air out of buildings with burning sulphur, followed by the wafting of more sweet-smelling smoke from the burning of camphor, orange leaves or sage. Other remedies, cited by Andrew Nikiforuk, in *The Fourth Horseman*, included wearing garlands of flowers (such as posies) to "purify the brain," and dietary advice:

> *"Eat figs with two filberts. Shun lettuce. Chew slowly and rise from the table hungry. Guard against weeping and fear 'for imagination works havoc with disease.' Light wine mixed with spring water [the original spritzer] keeps the body cool and free of disease."*[3]

Y. Pestis ignored all this, along with the prayers of priests, nuns and faithful congregations, and went right on a-killing, as had all of humankind's other invisible curses from time immemorial, from pneumonic plague to cholera, from bacteria-caused maladies to those caused by viruses or fungi.

Exactly why the Black Death broke out in such wild proportions, then subsided, provides a graphic illustration of systems theory in action. Once it reached Europe, the great plague was essentially a case of two systems colliding—system one being the life cycle of *Y. Pestis*, and two being the medieval human housing system. In the jargon of theorists, their collision provoked a series of "*positive feedback loops*," that is, effects which amplify an original stimulus, which then went rampant.

Y. Pestis is carried in the gut of the oriental rat flea (*Xenopsylla cheopis*), which can live on the bodies of both rats and humans. If the flea bites a human, it passes plague bacteria into the human victim's blood. European peasants built their homes without windows, partly because they couldn't afford glass and partly just to keep out the cold. Roofs were of thatched straw, walls of wattle, and the dirt floors were piled with rushes for insulation. The straw ceilings supplied perfect nesting for rats, from which fleas could drop down on the humans below (positive feedback loop #1). The rats, which liked shadows (loop

#2), also nested in the floor rushes (loop #3) and the wattle walls (loop #4), from which yet more fleas could emerge. Houses of the poor who lived in towns were packed tightly together, assuring easy passage for rats and *Y. Pestis*-infected fleas from one family to the next (loop #5).

As Nikiforuk, paraphrasing Philip Ziegler, notes: "A rodent urban planning committee couldn't have designed better rat accommodation."[4] All those positive feedback loops reinforced and amplified the "output" of the *Y. Pestis* life cycle, that output being human deaths, and would have gone on doing so forever unless some *negative feedback loop* (which curbs the original stimulus), or loops, intervened. The latter appeared when Europeans, goaded not by fleas but by fire, changed their preferred housing materials. Nikiforuk explains:

> *Plague visited London for the last time in 1665 because the Great Fire of 1666 destroyed 13,200 wood homes. In their place, Londoners constructed 9,000 brick houses with floor carpets and tile roofs. Similar fiery events transformed house architecture in Berlin, Brandenburg and other cities. While humans were making their dwellings less appealing to the black rat, the larger grey rat, Rattus norvegicus, began swimming in swarms across the Volga and displacing the black rat from all its traditional lodgings. Rattus norvegicus, a much wilder rodent, carried a different type of flea, eschewed humans and preferred sewers to houses. It didn't make plague-carrying a priority.*
>
> *Without proper housing, the plague retreated from Europe ...*[5]

Better housing served as a negative feedback loop, curtailing the reproductive spread of *Y. Pestis,* and its fatal output. Of course, the people of the day had no idea what had happened, certain only that they were glad the plague epidemics were over.

First steps

Science seems to advance in fits and starts, sometimes two steps forward and three steps back, but in the end it's a cumulative enterprise, each researcher building on the efforts of those who went before—that is, if he or she is aware of their work. Sometimes independent researchers find answers, unknown to each other, and only later do further investigators put the discoveries together.

The motives and attitudes of both scientists and laymen during this bumpy process have a direct bearing on its practical results—including the potentially apocalyptic ones we are witnessing today. The devil, as they say, is in the details. The development of "infection control" illustrates this.

The first solid step toward understanding what was happening to cause putrefaction and fevers came in 1546, when an Italian doctor, Girolamo Fracastoro, suggested epidemics were caused by seed-like entities transmitted by physical contact. Then, in 1590, Zacharias Janssen built what was probably the first microscope, and by the 1670s his fellow Dutchman, Anton van Leeuwenhoek, was actually observing a variety of microbes with such an instrument, and describing them (he wasn't able to see viruses, however, which are too small to be observed with the instruments of his day). In 1768 an Italian priest, Lazzaro Spallanzani, using a microscope, was able to demonstrate that microbes could be killed by an hour of boiling. It wasn't yet clear, however, that microbes caused diseases, and if so, which microbes or which diseases.

Spallanzani, who had already become a professor at the University of Reggio at the early age of 25, more than a decade before his discovery, was a scholarly man and teacher. While he won many honours and managed to earn a good salary most of his life, he was focused on knowledge, not profit.

The French chef and confectioner Nicolas Appert was a different story. In 1795, he began experimenting with food preservation, and eventually hit on what later became known as the "hot water bath" method of canning. He put food in glass jars, sealed the lids with cork and then boiled them in water. He found this prevented spoilage, but as far as is known, didn't really understand why. He may not have known of Spallanzani's work.

Appert was no academic, but a businessman, and was among the first to recognize the prospects for cash in the effort to fight putrefaction. The French military had offered a prize of 12,000 francs for a new method to preserve food, and in 1810, having perfected his hot water method, Appert submitted his idea.[6] He won the prize, along with a medal, and subsequently patented his invention. His Maison Appert, which he set up near Paris, became what may have been the first food bottling factory in the world, earning him, also, a good living.

The profit motive, though present in Appert's work, had no part in Spallanzani's. Science can advance with or without it.

In the 1850s, England's John Snow pioneered some of the ideas of epidemiology, showing that diseases spread from geographical point sources. Although personally sceptical of the "miasma" theory, he didn't know exactly how cholera, for example, actually spread from those point sources. Then, in the early 1860s, Louis Pasteur in France developed his "germ theory," demonstrating how several maladies were actually transmitted by microscopic organisms, and Robert Koch, researching anthrax and cholera, was able to show how specific pathogens caused specific diseases.

Pasteur also developed, in 1864, a method of wine and beer preservation, similar to Appert's, but using less extreme heat. He found one only had to heat wine to 50 or 60 degrees C to kill all the microbes in it and render it safe for storage. Later dubbed Pasteurization, the method was subsequently used to kill tuberculosis bacteria in milk.

Both Snow and Pasteur were scientists, pursuing knowledge. Neither, as far as we know, was primarily inspired by the prospect of getting rich.

Disinfectants/antiseptics

Harking back to Spallanzani's experiments, some 19th century surgeons had by this time begun routinely boiling their instruments (but not their patients) in water before operating. This process of *disinfection*, namely killing the microbes on non-living objects, was more effective than the old Roman method of passing instruments through flames.

Then, inspired by Pasteur's work, Joseph Lister, in his 1867 paper *Antiseptic Theory of the Practice of Surgery*, advocated the use of carbolic acid (phenol), a derivative of coal tar, as an *antiseptic*, to rid the surgical area and living skin surface of germs prior to surgery. The acid kills bacteria by first penetrating their cell walls, then initiating a precipitation effect (separation of a substance, in solid form, from a solution) on the cell's protoplasm. The protoplasm is changed to a granular solid, and the bacterial cell crumbles. Lister didn't know all these details, of course, but he could easily see that, applied topically, the acid decreased microbe-induced putrefaction.

Lister's idea caught on, but carbolic acid was irritating to the skin, and worse yet, was found to be highly toxic if used in too high concentrations, or in dressings left on the skin for longer periods. The carbolic from the dressings gradually penetrated the skin, killing the patient's own tissue and provoking gangrene. It was particularly corrosive to the eyes and respiratory tract.[7] Phenol vapour, if inhaled, also caused lung oedema. In addition, carbolic acid only killed some species of bacteria, leaving others undamaged.

Even before Pasteur's work and Lister's paper, other substances had been used to fight putrefaction. But their use was mostly empirical, that is, they were employed because they had the practical effect of preventing the problem, even though *how* they prevented it may not yet have been fully understood.

Among the early competitors of carbolic acid were iodine and alcohol. Iodine, a chemical element, was discovered by French chemist Bernard Courtois in 1811, and sparked unusual enthusiasm among early medical researchers, who began prescribing it for a wide—in many cases ridiculous—variety of illnesses. It was an early example of what has since been dubbed the *"Magic Bullet"*[8] belief, mentioned previously, which focuses on a single, supposed "miracle cure," and magnifies its powers, applying it everywhere. Iodine was prescribed for:

> *Paralysis, chorea, scrofula, lacrimal fistula, deafness, distortions of the spine, hip-joint disease, syphilis, acute inflammation, gout, gangrene, dropsy, carbuncles, whitlow, chilblains, burns, scalds, lupus, croup, catarrh, asthma, ulcers and bronchitis—to mention only a few.*[9]

Some of the applications, however, were genuinely effective, particularly Lugol's solution, first made by physician J.G.A. Lugol in 1829, and used subsequently as a topical antiseptic. Other iodine-based antiseptic formulae were also developed, and its dark purple or brown stain became a familiar sight in surgeries. Today, it is widely, and successfully, employed in solutions of povidone or sodium iodide.

Alcohol, usually a 70 per cent solution of ethanol, kills bacteria by "denaturing," or breaking the atomic bonds of their protein molecules, specifically their hydrogen bonds. When this happens, the molecules lose their shape and cannot function. However, alcohol evaporates from the skin fairly rapidly, leaving the surface open again to fresh bacterial invasion, and it doesn't usually kill bacteria in their dormant, spore phase. Ethanol is effective against numerous bacteria and some viruses, while the less-often-used isopropyl alcohol is slightly more effective against bacteria.

George Tichenor, a Confederate Army surgeon, experimented successfully with alcohol solutions on soldiers' wounds during the American Civil War, in 1863, and his work provides one of the earliest examples of a medical advance being used for self-interested ends. Some say outright cruel ends.

Tichenor first witnessed alcohol's antiseptic properties when he was himself severely wounded in the leg. Doctors advised amputation, but he treated the wound with an alcohol based solution of his own concoction and it healed. He then began using the solution, with similar results, on the wounds of his fellow Confederates—but, in what today might be called an act of germ warfare, gave strict instructions to those under him never to use it on wounded Union Army prisoners, who should be left to suffer and die.[10]

Tichenor, who was a businessman prior to the war, is also reported, like Appert, to have profited from his discovery, patenting it with the Confederate government. Afterward a picture of the Confederate Stars and Bars battle flag appeared on the label of the product for many years. This no doubt helped to sell it, down in Dixie.

The ethics of *patenting medical discoveries and making them available to sick people only if they could afford to pay—or were on the right side in a war*—had not yet become a major public issue. Up until at least the mid-1800s, discoveries were few and far between. And most were made freely available to the public, if only because there was no way to prevent it. Lister didn't "own" carbolic acid, for example, and couldn't stop other doctors from using it once his paper was published. Nor would he likely have wanted to. Even if he did, obtaining permission to use a patented item or process in those days was not all that expensive.

But the idea of using medicine as a weapon, or a way to get rich, pioneered by folks who thought like Appert or Tichenor, would eventually become commonplace.

Other familiar antiseptics include:

Merbromin, formerly sold under the trade name Mercurochrome. It is a mercuric salt, whose typical red stain was once a frequent sight on the cuts and scrapes of schoolyard children. First introduced in the U.S. around 1918, it was an effective topical antiseptic, quickly becoming ubiquitous. Skinned knees everywhere sported the familiar red smear, whose use bordered on Magic Bullet status. However, it was banned in the U.S. in 1998, and later in Germany and France, due to fears of mercury poisoning.

Hydrogen peroxide is also a powerful oxidizer that breaks down the molecules on the surface of bacteria, disrupting their cell walls and membranes, forcing the cell to split open and die. Unfortunately, it is such a strong oxidizer that it can damage skin or other tissue, causing scarring and longer healing times.

Boric acid in weak solutions has been used to treat yeast infections, as a wash for bacterial eye infections and as an anti-viral agent to diminish "cold sores" (caused by *herpes simplex labialis*, or less often *herpes zoster*).

Tea Tree Oil, an essential oil extracted from the leaves of the Australian Tea Tree (*Melaleuca alternifolia*), has only recently come into use as a topical disinfectant. Investigated by Arthur Penfold in the 1920s, who first reported its antiseptic properties, it is still insufficiently researched. Toxic if swallowed, it can also kill cats or dogs if applied topically in high doses. Research interest has been stimulated by its apparent effectiveness against antibiotic-resistant MRSA (for multi-resistant *Staphylococcus aureus*) bacteria.

There are many other antiseptics and disinfectants, from chlorine bleach to formaldehyde, as well as many other sterilization methods, too numerous to list here, but currently in use. Suffice it to mention only two more, very common, sterilization methods: steaming and radiation.

Autoclaves, which utilize steam under pressure (the same principle as the pressure cooker), are used to sterilize medical instruments. Autoclave treatment inactivates all fungi, bacteria, viruses and even bacterial spores. Unfortunately, one cannot autoclave living patients, or their body parts.

Radiation, which came into its own in the 20th century, is also used to sterilize both surgical instruments and manufactured food products. Non-ionizing radiation, in the form of ultraviolet light, is sometimes used to sterilize solid surfaces. Ionizing radiation, such as gamma rays or X-rays, penetrate below surfaces and are used in a variety of applications.

Like so many other medical tools, radiation also became a kind of Magic Bullet, with doctors prescribing X-rays almost willy-nilly, until the cancer-causing effects of low-level radiation were discovered and users became more cautious. Those effects were a clear example of two systems theory dicta: 1) "You can never do just one thing. In addition to the immediate effects of an action, there will always be other consequences," and 2) "Every solution creates new problems."[11]

First made known in pioneering studies by England's Dr. Alice Stewart in the late 1950s,[12] the dangers of low-level radiation were far from welcomed by the nuclear industry, government or the medical establishment, all of whom were heavily invested in radiation applications. Resistance to the evidence produced by Stewart and others who followed her, including Dr. Ernest Sternglass, of the University of Pittsburgh, continued well into the 1980s and 90s. Sternglass' evidence faced particularly strong resistance, including attacks on his professional competence and character.[13] For decades, money and militarism continued to trump human health, while thousands of people suffered and died.

Immunization

Though they might prevent local infections due to surgical or other wounds, antiseptics couldn't stop airborne or other diseases transmitted by breathing or swallowing, such as cholera, the dreaded Black Death, or the virus-caused smallpox.

Immunization against smallpox, at first via direct inoculation (referred to as "variolation") of the smallpox virus itself, was likely tried originally in China or India. Britain's Lady Mary Wortley Montagu, who encountered it while travelling in the East, described it in 1721 to Hans Sloane, the King's doctor. But it was an obviously dangerous technique, and remained mostly a curiosity in the West until the late 1700s, when several experimenters in England, Denmark and Germany (most notably British farmer Benjamin Jesty in 1774, and British surgeon Edward Jenner in 1796) observed the fact that farmhands—particularly milkmaids—rarely contracted smallpox, even when it was raging among the rest of the local population.

Those who were immune were those who had previously had cowpox, a similar but far less serious viral disease caught from dairy cows. No one knew it at the time, since microscopes powerful enough to show viruses hadn't even been invented, but cowpox and smallpox are closely related microbes, both belonging to the *Poxviridae* family and *Orthopoxvirus* genus. The two strains of smallpox, *variola major* and *variola minor*, share many characteristics with the cowpox virus (*Variolae vaccinae*) and its close relative (possibly its mutant descendant) the *vaccinia* virus.

Vaccination works by causing the vaccinated patient's immune response system to react to the invading microbe, creating *antibodies* (see below) to fight the disease. Since cowpox and smallpox microbes are so similar, the same antibodies are effective in killing both invaders. Of course, neither Jesty nor Jenner knew this, but they could see that, whatever the reason why, the method worked.

Jesty, though not himself a doctor, was evidently a bright chap, and worried that his family might not survive when a local outbreak of smallpox started to rage. He took a calculated risk, and inoculated his wife and two children by taking fluid from cowpox lesions and scratching it into their skins. They survived the epidemic, as did many others given similar treatment. But neither Jesty nor most of the other early experimenters made an effort to publicize their work, or to profit from it monetarily.

The publicizing was left to Jenner who, in a series of experiments in 1796, first drew pus from a milkmaid with cowpox and then scratched it into the arm of a young local boy. Six weeks later, in an act which today would probably get him jailed for reckless endangerment, he "variolated" the boy with live smallpox. History doesn't record whether the boy, no relative to Jenner, had much to say about this, or speculate what his family might have done to Jenner if the lad came down with smallpox and died. But luckily for Jenner the boy was now immune and didn't catch the fatal malady.

By 1798, Jenner had not only perfected his method but publicized it effectively, and "vaccination," the term he gave to the cowpox inoculation process, became common. Louis Pasteur in France further boosted the popularity of the method in the 1880s, when he developed vaccines against chicken cholera and anthrax, and later rabies.

Over the years, other researchers developed other vaccines, the most prolific inventor being the American microbiologist Maurice Hilleman. In a career stretching from the early 1940s to his death in 2005, most of it spent with Merck & Co., Hilleman and teams working under him developed more than 36 vaccines against both bacterial and viral pathogens, including measles, mumps, chicken pox, hepatitis A and B, meningitis, pneumonia and various forms of influenza.

(As will be discussed below, it is an open question whether a career like Hilleman's could be repeated in the private sector today, given the economics and current views of the pharmaceutical industry.)

Vaccination causes the human immune system to create antibodies. This can be done by injecting a patient with either dead or inactivated disease microbes, or with purified products derived from them. Some vaccines consist of killed bacteria or viruses. These are no longer dangerous because they are dead, but still provoke antibody production. Other vaccines contain living pathogens (disease-causing organisms),

living thing to kill another, unwanted one), fighting locust plagues in Mexico and Argentina with the deadly (to locusts) *Coccobacillus*. It was a harbinger of things to come, as well as an example of the systems theory rules "Nature knows best ... always try to find a natural solution to a problem, if possible," and "Don't fight positive feedback, support negative feedback instead. Don't poison pests, support their predators."[15] Systems theory hadn't been invented yet, and d'Herelle was thus way ahead of his time. His later work would follow the same principles.

By the time the First World War broke out, he was in France, working at the famed Pasteur Institute where he helped produce more than 12 million doses of medication, chiefly vaccines, for the Allied military. It was at the Institute, in 1917, that he first noticed what he concluded had to be "an invisible, antagonistic microbe of the dysentery bacillus."[16] The microbe was actually a virus, too small to be seen by microscopes of the era, which acted as a predator against bacteria, like a fox to so many rabbits. Its presence had been noticed before, by British researcher Frederick Twort, in 1915, but Twort made no effort to pursue his discovery.

Although viruses would not actually be seen until the 1930s, following invention of the electron microscope, d'Herelle was convinced they were living things, and targeted them specifically in a series of experiments. While analyzing the stools of dysentery patients, he found evidence that the invisible agent was not only alive, but attacked only certain bacteria. As Thomas Hausler recounts:

> *He reported that the phage was a living microbe because it constantly reproduced itself. One drop of a dissolved bacterial culture was sufficient to decimate a new culture within a matter of hours. This could be repeated as desired. The phage from dysentery patients did not grow on dead bacteria or other types of bacteria.*
>
> *The claim that his discovery represented a tiny microbe was bold, because he could have had a simple disinfectant substance in front of him. He offered evidence that a living organism was responsible for the effect by distributing a drop of a fresh mixture of a few phages and bacteria on a solid surface of nutritive gelatin. After 12*

> *hours, d'Herelle saw a so-called 'lawn' of bacteria with several holes. He concluded that individual bacteriophages had landed on these places and reproduced at the expense of the bacteria. D'Herelle argued that a chemical substance could never concentrate at one place. With this, he had also discovered the fundamental method for isolating phages that is still used today …*
>
> *D'Herelle stressed that the appearance of viruses in feces coincided with the recovery of dysentery patients. This led him to make another ambitious claim: his 'antagonistic microbe' triggered the cure of dysentery in patients.*[17]

D'Herelle called his newly-described creatures "bacteriophages," from bacterium and the Greek word for "eat," phagein.[18] Like his experiments using bacteria to kill locusts, his use of phages was a form of biocontrol, only this time using viruses to kill bacteria. It was also a classic example of a negative feedback loop, in which phages provided the negative feedback that reduced the output of pathogenic bacterial systems.

Despite a few ups and downs, d'Herelle's discovery caught on quickly, and between 1920 and 1930 phage therapy came to be used throughout western Europe and the U.S. Research gradually expanded knowledge of the viruses, confirming that, like predators elsewhere, certain phages target only specific kinds of victims. Lions eat zebras, and typhus phages kill typhus bacteria.

The exact mechanism by which they do so wasn't worked out until after the first phage was observed under an electron microscope by Helmut Ruska, in 1939. Phages look for all the world like tiny, upright moon-lander modules, with a large, angular head, a small neck, cylindrical main body, and a set of tail fibres arranged like insect legs. Killing is part of their reproductive cycle, as Hausler explains:

> *The reproductive cycle begins when the virus uses its tail fibres to attach itself to its victim. The details of what happens next vary according to the different phage types. But their aim is always the same: to get their genetic material, which is located in the head, inside the bacterium. T4, a well-studied phage infecting E. Coli,*

> *then contracts its tail sheath which pushes a tube located within the tail through the membrane of the bacterial cell. The phage's DNA is passed through the tube into the cell, where it takes control, brutally stops many of its vital functions and forces it to churn out new virus component—heads, tails, tail fibres—in production-line style. Then comes the final assembly. Finally, enzymes dissolve the wall of the bacterium from the inside and the newborn bacteriophages reach the exterior, ready to attack new victims.*[19]

Doctors in France, Germany, the U.S. and Latin America used phage therapy against a wide array of infectious diseases, from boils to cholera to staphylococcus and *E. Coli* infections. More than 10,000 vials of phages were produced in Brazil alone in 1925, and employed successfully against diseases like dysentery. "The dysentery phage is by far the best therapy for dysentery known to date," said Jose da Costa Cruz, of the Oswaldo Cruz Institute in Rio de Janiero.[20]

But phage therapy was not without its difficulties. Its very specificity, which targeted some bacteria while leaving others intact (including, as would become more significant today, the beneficial intestinal bacteria that aid human digestive processes), meant that batches of phages had first to be isolated, identified and then cultured in quantity before application to a patient, a time-consuming process. Only if phages were pre-tested in a tube with bacteria from the individual target patient could doctors be sure they had the right virus to cure that particular patient's infection. Also, in a phenomenon that foreshadowed far greater troubles in future, some bacteria managed to develop an immunity to some phages. To get around these problems, care givers began giving phage "cocktails," mixtures of several phage types or strains, which eliminated the resistance difficulty but further complicated the culturing stage, and eventually would lead to complications in the drug approval process in many countries.

A number of failures to cure were also reported, although most of these later turned out to be due to administration of the wrong phage, of phage mixtures that were too weak or insufficient in quantity, or of phages that had been weakened or killed by other substances. For example, a preservative packaged with some phages manufactured by the Eli Lilly company restricted the mixture's potency. Some batches were totally useless.[21] Stomach acid was also found to destroy orally administered phages. Doctors had to give patients sodium bicarbonate, a base solution, to neutralize stomach acids before phages were administered.

In the early days of phage use, when enthusiasm was high, patient records were often poorly kept, and testing for efficacy was haphazard and unscientific. Successes, though widely publicized, were sometimes statistically insignificant, and proper "double-blind" studies—in which neither doctors nor patients knew who was receiving a phage and who a placebo—were rare. Thus, even in their heyday, phages had their share of detractors and debunkers. The battle for acceptance seesawed back and forth, as a successful cure would first be ballyhooed by phage supporters, then downplayed or questioned by sceptics. D'Herelle himself, whose ego was far from modest, often fanned the flames of controversy by fiercely attacking opponents.

Nevertheless, whatever the drawbacks, phages remained the best treatment yet devised by humans to battle their deadly microbial foes.

Enter penicillin

The best, that is, until antibiotics, which entered the scene almost by accident, in 1928, with the discovery of penicillin by the Scottish researcher, Alexander Fleming.

Working in St. Mary's Hospital, London, Fleming noticed that a petri dish he was using to grow staph bacteria had been left open. The dish was invaded by a blue-green mould, around which all bacteria were killed. Analysis showed it was a *Penicillium* mould, later found to be *Penicillium notatum*. Fleming suspected he was onto something, but further experiments seemed to show penicillin couldn't live in humans long enough to kill bacteria. Dosage amounts later proved to be the problem.

It was left to Cecil George Paine of the Royal Infirmary in England, and other researchers, to perform the earliest successful human cures with penicillin. Paine is credited with the first recorded cure, in 1930, when he used penicillin to destroy gonococcal eye infections in five mid-1930s, several years after the advent of penicillin. Researchers at the German chemical firm IG Farben's subsidiary, Bayer AG, developed an antibacterial drug they called Prontosil. It turned out that the active agent in Prontosil was actually the sulfa molecule, on which the earlier patents had expired. The drug was thus available to any company, and Bayer's dreams of massive profits evaporated.

Sulfa drugs, nevertheless, were produced by numerous companies in Europe and the U.S. and notched a number of high-profile successes, curing life-threatening infections in both Winston Churchill and U.S. President Franklin Roosevelt's son, Franklin Jr. During the war, they were widely distributed in powder and tablet form in Western soldiers' first-aid kits, and proved effective against a wide range of bacteria, particularly streptococci. Films of the day often showed soldiers in the field sprinkling sulfa powder into wounds.

Phages, meanwhile, suffered repeated setbacks. In the U.S., controversy over phage therapy's effectiveness continued through the 1930s and into the early 40s, with poorly-designed tests and a still-incomplete understanding of how phages did their work providing critics with ammunition. The U.S. Committee on Medical Research (CMR) of the National Research Council (NRC) supported several well-designed studies, by René Dubos of Harvard University and others, that yielded promising results. But definitive tests on human subjects, planned for 1945, never got off the ground. By that time, penicillin and sulfa had already taken the lead and phages were effectively eclipsed.

The only significant Western wartime interest in phages came, ironically, from the German Army, which had been looking at a phage drug—polyfagin—developed by the German firm Behringwerke to cure dysentery. Polyfagin worked relatively well, and was used fairly extensively by the Wehrmacht in the field, particularly in Africa. But there was no careful testing to prove its effectiveness and after the war, with the success of the new antibiotics, it faded into disuse.

The rise to supremacy of antibiotics as the preferred defense against infections in the West also coincided with two fundamental socio/economic trends, which acted as catalysts to the process: 1) the general, across-the-board growth of corporate power, and 2) the concentration of the pharmaceutical industry, in particular, in the hands of fewer and larger corporations.

Corporate dominance

Corporations, loosely defined as entities made up of many people *acting together as one* for business or legal purposes, have been around for at least 2,000 or more years, with examples in both ancient Rome and India. Most of the earlier ones were chartered by governments, or set up by already existing social groups, such as individual towns or cities, or the medieval craft guilds. Purely private, commercial corporations, set up for the express purpose of making money, came on the scene in the 18th century, and grew gradually more powerful and independent over the next 200 years, thanks to key enabling legislation in England, the U.S. and elsewhere.

Later chapters will go into this history in more detail, showing the effects of corporatization in sector after sector of society. Suffice it to say here, however, that the corporate business model had already become prominent by the 1930s, and in the post-Second World War era began a concerted drive to dominance. As one *Adbusters* writer put it:

> *They merged, consolidated, restructured and metamorphosed into ever larger and more complex units of resource extraction, production, distribution and marketing, to the point where many of them became economically more powerful than many countries. In 1997, 51 of the world's 100 largest economies were corporations, not countries. The top 500 corporations controlled 42 per cent of the world's wealth.*[23]

The pharmaceutical industry in particular also became corporatized in the postwar years, but remained, at least by today's multinational-conglomerate standards:

> *... relatively small-scale until the 1970s, when it began to expand at a greater rate. Legislation–allowing for strong patents, to cover both the process of manufacture and the specific products–came into*

didn't go into sufficient production to reach civilians until 1949.[27] Partly out of national pride—government propaganda of the time "painted pharmaceuticals, which were being developed and promoted by the West, as something foreign and suspect"[28]—and partly out of necessity, phage research continued, in Tbilisi and elsewhere, as did phage use by physicians. Anna Kuchment reports:

> *[Chief Surgeon Ruben Kazaryan] says, 'I use [antibiotics] only in life-or-death cases where there is no other choice. When you explode an antibiotic in your system, it kills everything around it, both bad and good. Antibiotics are immuno-depressing, because they work instead of your body.' Kazaryan's belief is that phages help a person's natural immune system to conquer an infection on its own. 'Phages decrease the number of bacteria to the point where the immune system can finish the job by itself,' he says.*[29]

Thus, as the *Wikipedia* entry for "Phage Therapy" recounts: "Russian researchers continued to develop and to refine their treatments and to publish their research and results. However, due to the scientific barriers of the Cold War, this knowledge was not translated and did not proliferate across the world."[30]

Soviet scientists could obtain permission to travel outside the USSR only with great difficulty, making their attendance at medical conferences and symposiums nearly impossible. Few Western researchers could read the Cyrillic alphabet, and the university or other employers of the few who could were likely not inclined to credit the Cold War arch enemy with anything. Thus, Western advances in infection control slowly made their way East, but it was a one-way highway, with nothing coming back.

Not until Gorbachev's *glasnost* opened up the USSR, and the Soviet empire crumbled, did things change. By then, the problem of bacterial resistance to antibiotics had begun to worry Western doctors, and eventually pressure built up to look for alternatives. Unfortunately, the collapse of the eastern empire caused major economic dislocations within its former member states, and Moscow's centralized government support for phage research and development dried up. The West's belated revival of interest in phages had to focus on a few former republics, such as Georgia, where treatment and research organizations like the Eliava Institute, mentioned above, were now struggling.

Zombie bugs

At first, the pilgrimage eastward was little more than a trickle, made up mostly of either well-to-do or desperate patients who'd run into one or another newly-resistant microbe and found their situations grave enough to warrant the expense of travelling abroad for relief. But, as more and more microbes became resistant to more and more antibiotics, like tiny zombies that would not die, the number of eastward-bound travellers increased, and western researchers' interest in supposedly archaic phage therapy began to rise again.

Before looking at this beginning of a revival, however, the reasons why antibiotic resistance became a problem in the West need to be understood. In many cases, they were the same reasons that originally made antibiotics desirable: they were the Magic Bullet, fast and easy to use, and effective across a broad swath of microbial species. Doctors came to prescribe them almost automatically for any infection, even a cold in the nose, and when they didn't volunteer to do so, patients indignantly demanded them. Patients even insisted on them for cases of influenza, which is a viral disease and not treatable with anti-bacterial medicine. Antibiotics were everywhere.

All too often, patients wouldn't use all the pills prescribed, but would stop the course of treatment as soon as the most obvious symptoms disappeared, and flush the surplus tablets down the toilet. The result was that some residual bacteria in the patient weren't killed, and developed immunity, as did the microbes in thousands of sink and toilet drains, and in the sewage pipes that emptied into nearby rivers, and eventually the water supplies of people downstream. The more doctors and the public got used to antibiotics, and came to take them for granted, the more common such careless practices became.

But the worst, industrial-scale source of resistance was the livestock food industry, where dairy and beef, pig, chicken and other producers had for years known that even minor illness in their animals could lead to loss of production. Sick dairy cows, even those only suffering from the cow equivalent of the common cold, produce slightly less milk than healthy ones, and sick beef cattle, hogs and chickens put on less weight, because a percentage of their metabolic energy is being used to fight off the infection, rather than make milk or meat. Veterinarians prescribed antibiotics for their patients to remedy this, which, since the animals really were sick, was perfectly proper.

However, it wasn't long before farmers, rather than wait for a cow or pig to actually fall ill and need treatment, began insisting on "preventive" medicine. They wanted antibiotics made part of their animals' regular feed ration, even when they weren't sick. Soon, surveys were showing that production increases occurred in herds or flocks fed antibiotics as part of their regular rations, whether or not any infection had ever appeared or even threatened to appear. Antibiotics became the new Magic Bullet for meat and milk production, essential for boosting profits.

They became particularly important on industrial, or factory farms, where massive numbers of cattle may be jammed into feedlots to fatten just before slaughter, or thousands of so-called battery chickens, or hogs, may be crammed together in tiny, torture-chamber cages and pens, and where any disease, once started, could not help but run rampant. It was the story of the Black Death being written all over again, *with animal housing, rather than medieval human habitations, creating a potential positive feedback loop for the multiplication of disease microbes.* All that was needed was a trigger.

Back to bite

This author once worked for the UN Food and Agriculture Organization (FAO), and as far back as the late 1980s can recall staff commenting on the practice of "preventive dosage." One experienced immunologist once told me, over lunch in the staff cafeteria: "Mark my words, this is going to come back and bite us." And so it has.

As far back as 2000, a World Health Organization (WHO) report on infectious diseases warned:

> *Since the discovery of the growth-promoting and disease-fighting capabilities of antibiotics, farmers, fish-farmers and livestock producers have used antibiotics in everything from apples to aquaculture. This ongoing and often low-level dosing for growth and prophylaxis inevitably results in the development of resistance in bacteria in or near livestock, and also heightens fear of new resistant strains 'jumping' between species.*[31]

Systems theory warns that "everything is connected to everything else," and that "obvious solutions can do more harm than good,"[32] but systems principles don't have a place in factory farm planning. The WHO warning was ignored, by both the corporations that oversee and benefit most from the industrial farm system, and the governments their lobbyists control. It has been ignored so completely that by now, as Robert Lawrence, director of the Center for a Liveable Future at the Johns Hopkins Bloomberg School of Public Health, reports:

> *[U.S.] Food and Drug Administration (FDA) data have shown that 80 per cent of antibiotics sold in the U.S. are sold for use in food animals, not humans. The same data suggest that the vast majority of antibiotics used in food animals are administered to compensate for crowded and unsanitary conditions and to speed animal growth. This use promotes the development of antibiotic-resistant bacteria that can spread to humans on food and through the environment.*[33]

Eighty per cent.

Lawrence notes that industry apologists object that some antibiotics used this way are not the same as those prescribed for human patients. He points out, however, that many of the latter belong to classes that can affect resistance to other, human-use antibiotics. Hausler agrees, and gives an example:

Drug discovery ... has been driven by profit signals in the commercial marketplace. Because the easy to find antibiotic chemical compounds have been discovered, new research is complex and expensive—yet antibiotics are priced as low-margin commodity drugs. Worse, regulators—especially the FDA in the U.S. which, as the gatekeepers to the richest drugs market in the world, is of pivotal importance—only licenses antibiotics for particular rather than generalised uses.

For a profit-minded drug company there is an obvious conclusion. Don't waste valuable R&D on low-margin antibiotics that are never going to become super-profitable blockbusters. Instead, develop, say, Viagra ... which is a high-margin, global money spinner.[38]

With only two exceptions, no new classes of antibiotics have been discovered since the 1980s.[39] S.J. Projan and D.M. Schlaes summarize what's happened during those years:

In the early 1980s, the pharmaceutical industry began scaling back on their antibacterial drug discovery efforts, with approximately half of large US and Japanese pharmaceutical companies ending or curtailing their efforts. Yet antibacterial drug discovery efforts did continue at many major US and European pharmaceutical companies through the 1990s and these efforts led to the introduction of quinupristin-dalfopristin (Synercid) and linezolid (Zyvox), both targeting Gram-positive pathogens, to the marketplace in 1999 and 2001 ... But since 1999 the industry has once again pulled back from anti-infective research in an even more concerted manner, with 10 of the 15 largest companies ending or curtailing their discovery efforts.[40]

A complicating factor, industry consolidation, made things still worse:

While this [cessation of research] was occurring the industry has been experiencing a series of mega-mergers leading to large scale consolidation. Five modern pharmaceutical companies (Pfizer, GlaxoSmithKline, Novartis, Bristol-Myers Squibb and Aventis) actually are comprised of 32 progenitor companies, all in business as

> *recently as 1980. Many of those companies had their own antibacterial groups, so consolidation alone has resulted in a major decrease in the hunt for novel antibacterial agents.*[41]

In short, there are fewer companies and fewer researchers engaged in drug development generally, and, thanks to the effects of the negative feedback loops just mentioned, it is plainly more profitable for those who are still so engaged to help older men get erections than to save their lives if they are dying from bacterial infection. The search for new antibiotics has all but ceased.

The profit motive, or rather the lack-of-sufficient-profit motive, is also the main reason why there is not yet, at this writing, a proven effective vaccine against the lethal Ebola virus. Some promising possibilities exist, but none are in production. Whether the enemy is bacterial or viral, the same rules apply. The director of the World Health Organization, Margaret Chan, put it bluntly: "A profit-driven industry does not invest in products for markets [namely sub-Saharan Africa] that cannot pay."[42]

Noting that Ebola first surfaced in the Congo as long ago as 1976, and that there have been 22 outbreaks since, none of which prompted Big Pharma to spring into action, Erica Etelson observes that "had Ebola landed on US shores sooner, a vaccine would already be available ... the lack of an Ebola vaccine—and the wildfire spread of the virus—are a direct result of private sector control over vaccine development and the absence of public health infrastructure that could have contained the outbreak ... If market research projects insufficient demand for a product or service to meet corporate goals, that product or service doesn't come to be, regardless of the fact that it's vital for the health of people or the environment."[43]

Alarmed at the overall situation, the WHO released a report in 2014, warning that drug-resistant "superbugs" have spread to every corner of the world. "Increasingly, governments around the world are beginning to pay attention to a problem so serious that it threatens the achievements of modern medicine," said Dr. Keiji Fukuda, the WHO assistant director-general for health security, in the report's foreword.

"A post-antibiotic era—in which common infections and minor injuries can kill—far from being an apocalyptic fantasy is instead a very real possibility for the 21st century."[44]

Meanwhile, back in Georgia

Meanwhile, that once, and potentially future, rival of antibiotics, bacteriophage therapy, waits in the wings, hoping for a comeback. But the wait may prove a long one.

Back in the former Soviet Republic of Georgia, the Eliava Institute still struggles to keep its doors open and stay in the black, despite a few recent flurries of Western interest in its experience with phages. A post to the Phage Therapy Center website recalls its glory days:

> *In its heyday in the 1970s and 80s, nearly 800 people worked in the Industrial Branch of the Eliava Institute, using enormous vats, pill stampers and automatic bottling machines to pump out tons of phage products for military and civilian uses all over the Soviet Union. Another 200 worked to analyze hundreds of thousands of bacterial samples that continuously poured in at the direction of the Soviet Ministry of Health, testing the phage cocktails for efficacy and constantly isolating new phage and making refinements. They also fought infectious disease in other ways—vaccines, immune enhancers, probiotic bacterial cultures—but phage were their main focus. By then, institutes and factories in places like Gorki and Ufa were also producing these phage products for Soviet use, but Tbilisi phage were especially prized as far away as Lithuania even in 1990.*[45]

Then came Gorbachev's *perestroika* and *glasnost*, and the fall of the Soviet Empire. Georgia declared its independence from Moscow, and Russia stopped all funding to the institute. Demand for phages from other republics and the Red Army ceased. "The mighty production facilities were broken up and privatized, and [their] ties to the institute cut," writes Hausler. "The researchers were left behind in their labs, without salaries or research money. Meanwhile, production dropped to a minimum."[46]

Elizabeth Kutter, an American professor at Evergreen State College in Olympia, Washington, whose specialty was phage biology, visited Georgia in 1990, and stumbled upon the institute. She contacted journalists in the U.S. and alerted them to the story. An article on the institute in *Discover* magazine (by Peter Radetsky, in 1996), as well as mentions in other publications, attracted some attention. A number of American and other Western patients, suffering from seemingly incurable infections by antibiotic-resistant bacteria, travelled to Georgia and were successfully treated there.

Several individuals, including Michel Chretien, the physician brother of former Canadian Prime Minister Jean Chretien, joined efforts to launch businesses in the West to utilize phage therapy, and a number of companies were founded. Some Georgians also migrated west, such as Dr. Alexander Sulakvelidze, former director of molecular microbiology for the Georgian National Center for Disease Control, in Tbilisi, a well-known phage researcher who helped set up Intralytix, Inc., in Baltimore, Maryland, where he is currently vice president for research.

But the initial surge of interest in the human health applications of phages never hit high gear. Several startup companies disappeared, and others have seen only modest growth. GangaGen, a company which originally had both Canadian and Indian branches, and for which Dr. Chretien served as a scientific advisor, has since closed its Canadian branch. Chretien, who in 2008 was giving enthusiastic public talks about the life of Felix d'Herelle,[47] now rarely discusses the subject.

A list of drawbacks have crippled growth. Some are unique to phages, and some are similar to those that currently hamper new antibiotic development.

One key drawback has been the fact that bacteriophages are, in themselves, difficult or impossible to patent. A company that develops a chemical rat poison like, say, warfarin, can patent the compound and, at least for the duration of the original patent, retain the exclusive right to sell it at whatever price it thinks the market will bear. But no company can patent house cats, or owls, or any of the other living predators that attack rats. They are living creatures, and as such unpatentable.

Similarly, antibiotics are biochemical "poisons" that kill bacteria.

Once developed, they can be patented. Phages, in contrast, are living predators, creatures which have existed on planet Earth for literally billions of years, far longer than housecats! No one can patent them, particularly in the U.S., where several recent Supreme Court decisions, notably *Mayo vs Prometheus* (2012) and *AMP vs Myriad* (2013) have made it particularly difficult. As Timo Minssen reports:

> *In its unanimous Prometheus opinion, the Supreme Court held that a claim to a 'natural law' is not patentable unless it has additional features that add significantly more to the 'natural one' itself ... Then, in Myriad, the court unanimously held that 'a naturally occurring DNA segment is a 'product of nature' and not patent eligible merely because it has been isolated.*[48]

Nor can they easily claim patent rights on combinations, or cocktails, of bacteriophages, developed by researchers to target certain bacteria and overcome the potential problem of resistance. As Karl Thiel, quoting Elizabeth Kutter, notes, the Georgians "almost always use cocktails. A typical, off-the-shelf phage therapy for purulent infections might include 30 phage strains going after five different types of bacteria. And that preparation may vary from hospital to hospital ... and, because the concept of using phages [in cocktails] as therapeutics is almost a hundred years old, it is unpatentable."[49]

And even if "entrepreneurs could secure protection for an individual phage strain they have characterized, with an estimated 10 to the 8th power strains of phage in the biosphere, there's nothing to stop a would-be competitor from finding a different but perhaps equally effective strain."[50] Persian cats are as good at killing rats as Siamese, after all, and as easily found.

"Few companies would be willing to invest the sums of money required to develop a product for bacteriophage therapy unless their results are protected by an international patent," concludes the University of Brighton's Jonathan Caplin.[51]

Despite such handicaps, a handful of European and North American companies have obtained patents, not on the phages themselves but on the ways they are applied, or the combinations in which they are applied. For the most part, however, their focus is not on human health applications, but veterinary or food safety uses (more on this below).

Regulatory hurdles

As noted previously, development costs for any new "drug" can be substantial, possibly topping $500 million. They run particularly high in North America, where the regulatory hurdles to the introduction of new drugs for use in human subjects are formidable—a negative feedback loop, with a capital N, for any drug production system. Those hurdles were definitely not set up for phages to leap easily over. In fact, they show a fundamental lack of awareness of how phage therapy works.

Dr. Steven J. Projan, a vice-president at Maryland-based MedImmune LLC, is a sceptic where phage therapy is concerned, and believes its revival would be financially impractical, at least in North America. In a 2004 paper that drew considerable critical reaction, he asserted that "phages can be teachers, technicians and tools but, alas, probably not therapies."[52] Reached by phone, he added:

> *The main cost is drug approval, going through the clinical trials. It's hard to get [U.S.] regulators to be flexible in terms of the size and scope of trials. In the U.S. it's still a very conservative undertaking.*
>
> *In Europe, the European Medicines Agency (EMA) is showing more flexibility, with the EU Innovative Medicines Initiative. If I set up a clinical trial in Europe, the EU will directly pay the European clinical investigators to perform the study. In the U.S., the National Institutes of Health (NIH) tends not to fund studies on innovative products.*

The *Wikipedia* entry for Phage Therapy echoes Projan's position, explaining:

> *Due to the [target bacteria] specificity of individual phages, for a high chance of success a mixture of phages is often applied. This means 'banks' containing many different phages must be kept and*

Efforts to develop phage spinoffs, nevertheless, seem a bit contrived, as if pest controllers, rather than trying to get approval from a cat-sceptical agency to use cats to kill rats, decided to get approval to use cats' claws, or cat's teeth, but not the cats themselves!

As for agriculture and veterinary medicine, where regulatory approval is less difficult, several companies are now involved. Food safety measures are an example.

"Intralytix is concentrating on *Salmonella* and *Listeria*," writes Thomas Hausler.[59] "While *Salmonella* like to infest chickens, they don't make them sick. *Listeria* are found on meat, fruit, vegetables and certain types of cheese. Battery egg farms are breeding grounds for *Salmonella* because of the cramped conditions in the henhouses. The shells of freshly laid eggs are often infested with these bacteria ... The Intralytix researchers have found a few stages in the industrial life of fattening chickens during which they can use phages to reduce the danger of *Salmonella*. When they spray freshly laid eggs with their cocktail of viruses from the water of Baltimore's harbour, they can lessen the number of *Salmonella* on the eggs 1,000 times over."

Another company, PhageBiotech, which got out of human medicine, is developing phages to keep breeding shrimp disease-free.[60]

Dr. Rosemonde Mandeville,[61] former CEO of the Montreal-based Biophage Pharma Inc., which was originally focused on human phage therapy, now heads a different firm, Phagelux, with different objectives. Her prognosis for the North American situation is based on experience:

> *There is a need. Antibiotic resistance is increasing, there are no new antibiotics in the pipeline, and phage therapy is there. It's been there since 1917 or so. It's not a question of patents. You can have patents. You can [even] apply for a cocktail, if each phage of the cocktail has been well identified and produced separately, and shown that you can produce it in a consistent way. They require you to treat it phage by phage, and show when you have the cocktail they are there and the proportion is one-to-one.*

She adds, however, that having a patent isn't enough. For regulatory approval to use phages in human therapy "you have to do clinical studies, randomized clinical studies,[62] and it costs a lot of money." At Biophage Pharma, "we were focused on human therapy, on staph (*staphylococcus*) because of the problem of MRSA. I've tried for years and years to find a partner and find money to start doing something with that, as far as clinical studies are concerned, and there's no money available, government [research] money, money from big funds."

Only the mega-corporations of Big Pharma could find the cash, she believes. But they appear to see no short-term incentive since the profit margins would be too low after investment, compared to those for something like, say, Viagra. They do, nevertheless, seem to have a longer-term interest in phages.

"Each one of them, I know for sure, has a division of phage therapy," Mandeville said. "All of them are working on it, as an adjunct to antibiotics. When antibiotics alone don't work, you can give antibiotics plus phages. This is how it's going to work out. [But] human application is not interesting for them [today], because it's going to take a billion dollars to do that. It's not the human application they want. What is important for them is the bottom line. Are we going to make money fast, as fast as possible? I don't think they want to save the planet. This is not their problem. They want to make money. So in their minds, it's how are we going to make money quick. And the only way you're going to make money quick is food safety and animal health. It's a quarter of the price [for regulatory studies] to do a spray for tomatoes, or for hides of beef, than to do it for humans."

The demands of the FDA and other North American regulators for randomized clinical trials, or double-blind studies proving the effectiveness of phage treatment appears, at least on its face, parochial. There have been numerous randomized studies of this kind, notably in Poland, where several have been supervised by the Polish Academy of Sciences.[63] And the experience of other eastern European countries, including Russia during and after the Second World War, includes successful phage treatment of tens of thousands of soldiers and civilians. But, possibly as a subliminal legacy of the Cold War, work done in Eastern Europe tends to be denigrated in the West.

"There were some big preventive studies in Poland," Mandeville said. "They took a street and one side of the street got phages and the other did not, and they looked at the occurrence of dysentery in children. The ones that got phages had no dysentery. [In another study] they have treated 2,500 soldiers with gangrene and saved them. We're talking about big numbers. But the FDA will say it's anecdotal." Are North Americans really saying it must be 'our randomized study, not somebody else's'? "Absolutely," Mandeville replied. "It must be my [meaning a western bloc] thing, mine." To break down the barriers, "you need a pivotal study," she added, a western study that is so authoritative its results cannot reasonably be denied.

"And a pivotal study is being done."

That study, dubbed the "Phagoburn" project, and launched in 2013, is being conducted on burn patients in three Western European countries: France, Belgium and Switzerland, as a European Research and Development project funded by the European Commission. The study, involving multiple hospitals in all three countries, "aims at evaluating phage therapy for the treatment of burn wounds infected with the bacteria *Escherichia coli* and *Pseudomonas aeruginosa* ... through the implementation of a phase I - II clinical trial."[64]

Will it put paid to FDA and other regulators' reluctance? Perhaps, perhaps not.

But if Big Pharma finds a way to make phages, or some patentable phage spinoff, more profitable than they look now, some observers think regulatory approval would follow quickly, even without Phagoburn. This view may be overly cynical, but the pharmaceutical majors unquestionably have a great deal of political muscle, as the U.S.-based Center for Public Integrity reported in 2005:

> *The pharmaceutical and health products industry has spent more than $800 million in federal lobbying and campaign donations at the federal and state levels in the past seven years ... Its lobbying operation, on which it reports spending more than $675 million, is the biggest in the nation. No other industry has spent more money to sway public policy in that period. Its combined political outlays on lobbying and campaign contributions is topped only by the insurance industry ...*

The industry's multi-faceted influence campaign has also led to a more industry-friendly regulatory policy at the Food and Drug Administration (FDA), the agency that approves its products for sale and most directly oversees drug makers ... Medicine makers hired about 3,000 lobbyists, more than a third of them former federal officials, to advance their interests before the House, the Senate, the FDA, the Department of Health and Human Services, and other executive branch offices.[65]

More recently, *Forbes Magazine* published an analysis by BioMed-Tracker:

As recently as 2008, companies filing applications to sell never-before-marketed drugs, which are referred to by the FDA as 'new molecular entities,' faced rejection 66 per cent of the time. Yet so far this year the FDA has rejected only three uses for new chemical entities, and approved 25, an approval rate of 89 per cent. But, Forbes added, the 2015 rejection count includes rejections of Avycaz, a new antibiotic from Allergan, for hospital-acquired pneumonia, and selling Jardiance, a diabetes drug from Eli Lilly and Boehringer Ingelheim, in combination of metaformin. But Avycaz was approved for two other uses and Jardiance is on the market by itself. So in reality, the FDA approval rate is more like 96 per cent.[66]

As long ago as 1969, Dr. Herbert L. Ley Jr., on retiring from the post of Commissioner of the FDA, told the *New York Times*: "The thing that bugs me is that the people think the FDA is protecting them—it isn't. What the FDA is doing and what the public thinks it's doing are as different as night and day." He added that there was "constant, tremendous, sometimes unmerciful pressure" from the drug industry and that drug company lobbyists and politicians brought "tremendous pressure" on him and his staff to prevent FDA restrictions on their drugs.[67]

So, is the FDA too strict, in not granting phage therapy approval? Is it too lax, in granting approval to 96 per cent of other "drugs"? Is it simply waiting for directions from Big Pharma to approve or disapprove and which eventually grew resistant to antibiotics. "I have a 38-year history of this," she said. "I was looking over my income tax returns recently and in medical expenses there were 15 prescriptions for antibiotics in one year. One of the last infections I had, an *e. coli*, when tested was resistant to 15 out of 20 possible antibiotics, so I was in real trouble. I had a urologist who was treating me [and] an infectious disease specialist, who told me there was [so much] antibiotic resistance we had better stop treating [with them]. They were also using fluoroquinolones, which were horrible—the pain. I had taken them over and over with no seeming problem, until there was a problem, and then it was cardiac arrhythmia, and severe tendinitis. A horror show. My life was becoming more and more one of isolation, pain and sickness, and it was awful. I was afraid to walk down the street in case I had a heart arrhythmia where you pass out. So I went online and started digging."

Eventually, she found references to bacteriophages, and to the Eliava Institute in Tbilisi. She also found "a study done here in Canada, a collaboration between the University of Guelph and McGill University, saying phage did show promise against *e. coli*, and I said OK, what have I got to lose?" The *e. coli* weren't the only bacteria attacking her. "When I went to Tbilisi, the main ones were *Enterococcus faecalis* and *e. coli*. I had staph, *staphyllococcus aureus*, and in Tbilisi they discovered *Klebsiella pneumoniae*.

"Travelling to Georgia seemed crazy for us. It was so far. But at that point it was, if we have to sell the house, we're going. We just kept putting money out, because I had no more life left. I ended up being there two months. If you don't have money, you can't do it." In Tbilisi, she said, there were many tests, and treatment with phages was supplemented by vitamin and other support therapy. "They look at the whole person."

It worked.

"When I came back, I went from being unable to participate in anything, to being able to visit a local farm with my granddaughter. We had fun. In Tbilisi we saw patients from France, Germany, going in looking pretty serious, and coming out improved, all smiles."

2) ***Treatment by a licensed physician and/or nurse, who imports phages from abroad to employ here***.

Several U.S. doctors have taken this route for patients in dire need, and at least two MDs, in Texas and Washington State, have done so publicly.

Dr. Emily Darby, a Seattle, Washington infectious disease specialist, had been treating Rachel George, a young woman born with a chromosomal abnormality that made her prone to a variety of disabilities, including vulnerability to infections. "Kidney infections, urinary infections, heart infection, brain infection, bone infection, you name it, she's figured out a way to get it," her mother, Rose George, told a local paper:[73]

> *The worst of it came about two years ago, when two infections—MRSA [multi-resistant staphylococcus aureus] and pseudomonas—settled in Rachel's lungs and wouldn't let go. They tried virtually every antibiotic available—but Rachel's infections were drug resistant.*
>
> *Doctors prepared the family for Rachel's death.*
>
> *"It's horrible. It's frightening and you become very desperate," says Rose. "At that point, the nurse that did the home visits said, 'Have you ever heard of phage therapy?'"*[74]

Dr. Darby decided to take a chance on what seemed a long shot, partly because she was aware that phages are target-specific, attacking only the bacteria on which they regularly prey. "One of the reasons I said I think it's fine to go ahead and give it a try is, it's not likely to be harmful, and she was in such a bad position that we were willing to try almost anything to help her get better," she said.

In a separate interview, Darby noted the hospital was not as willing to make the attempt, and Rachel had to be treated at home. "The inpatient facility here refused to allow phage treatment because of concern for perceived infection control risk, which I imagine would be a barrier with other institutions as well."[75] A phage is, after all, a virus, and so carries its own aura of fear of the unknown. Nor did insurers cover the treatment, forcing the family to pay up to $800 for a three-month supply of phages.

But the results were spectacular:

> *"From the first phage treatment that we did, the MRSA disappeared. And we'd been battling MRSA for almost three years at the point we started this, and it was gone," says Rose George.*
>
> *The MRSA cleared up. Then the pseudomonas infection went away. And Rachel changed.*
>
> *"It's not just that Rachel's not getting as sick anymore; it's that she's thriving," says Rose. "Horseback riding, cheerleading, going to the mall, blowing kisses to guys ..." When Rachel was born, her family didn't know how long they would have her. She is now 31 years old—and this month she celebrated a remarkable anniversary.*
>
> *With phage therapy, she's gone an entire year without being admitted to the hospital. Rachel's family says she's not just surviving. She's living.*[76]

Despite such positive results, Dr. Darby doesn't foresee many similar cases in the near future.

> *"I have not prescribed phage therapy to other patients," she said.*[77] *"It has not been rigorously scientifically studied, so it is hard to really recommend to patients. I think it should be studied further though. The major issue is going to really be financial/practical. Phages are available in the environment, so not really patentable unless in a specific mixture. Drug companies are not going to want to test proof of principal with patient-specific phage therapy (matched from environmental samples as they do in Tbilisi), but rather a generic mixture ... No patent potential, in essence, means no study."*

Meanwhile, at the Lubbock, Texas, Regional Wound Care Center, Dr. Randall Wolcott has also used phage therapy, with considerable success. One of his patients, Roy Brillon, suffered from surface *streptococcus* and *staphylococcus* infections on his thighs that ate away the soft body tissue and refused to respond to antibiotic treatment. Extremely painful, the lesions eventually forced him to quit his job as a house painter. Dr. Wolcott prescribed morphine for the pain.

"I was only supposed to take two pills a day, but I was taking three in the morning and three in the afternoon," Brillon said. "The pain is indescribable. You just grit your teeth." A report in *Popular Science* added:

> *Wolcott knew well the typical prognosis for patients with antibiotic-resistant infections like Brillon's: gangrene, amputation and, for about 100,000 Americans a year, death. "Chronic wound is a code word for you can't heal it," [Wolcott said]. "The hallmark is, we cut it off or we cut it out. It's pretty barbaric." Wolcott was desperate for an alternative. After putting in 10-hour days at the clinic, he often sat up late at night poring over medical journals for the newest wound care research—something, anything that might help patients with the most intractable infections.*[78]

Eventually, he found phages.

> *Wolcott had to petition his state regulatory board to allow him to administer [phage] only to people who had exhausted all other options. Then because you can't find phages in U.S. pharmacies, he had to trek all the way to the former Soviet republic of Georgia to get it. There it's sold over the counter like eyedrops. He bought, for US$2 each, three clear glass bottles, each filled with a liquid containing hundreds of types of phages.*[79]

He began dribbling drops of the liquid into Brillon's wound, and within three weeks it was completely healed. "You'd better take pictures of this," Brillon told Wolcott, "or nobody is going to believe it."

Since then, Wolcott has treated several other patients with phage solutions, and they showed similar results.

He hasn't encountered any opposition from state authorities. "I told the Texas Board of Examiners, my licensing board, I wanted to use bacteriophage as a complementary or alternative medicine, like aloe for burns. That's a plant, a natural product, and phages are natural. The federal Complementary and Alternative Medicine Act, the CAM Act, also called the Hatch Act, basically says if you want to use something natural, to settle your stomach or treat your rash or your wound, that's legal. And physicians can do it as an adjunct to, but not in place of, standard care. Bacteriophages do not preclude us from using the regular antimicrobials, biocides, antiseptics, antibiotics. But in high risk legs, if there's *staph* or *pseudomonas*, why not seed it with these phages which have been shown to propagate and kill that bacteria?"

In Canada, provincial licensing bodies like the Ontario College of Physicians, "don't comment on any type of treatment," according to media spokesperson Prithi Yelaja, but are unlikely to intervene unless asked. "If we receive a complaint [from a patient or physician] we will investigate, but we don't comment to media on any treatment," said Yelaja. The Royal College of Physicians, which is a national body, "defers to the provinces," while the Canadian Medical Association (CMA), has no regulatory role, noting "it [phage therapy] would definitely be a College issue." Media spokespersons for the American Medical Association (AMA) had not heard of phage therapy, and also noted it would be a matter for state licensing bodies to regulate.

Wolcott has reported his experiences in peer-reviewed medical journals, as well as in the popular press. He co-authored a 2009 *Journal of Wound Care* paper on a safety trial of phages for treatment of venous leg ulcers, using a phage cocktail (WPP-201).[80] The paper concluded:

> *This phase I trial demonstrated the safety of WPP-201 in participants with chronic wounds ... its efficacy in managing infected wounds remains to be demonstrated by rigorously designed clinical trials. However, given the current difficulties encountered with managing chronic wounds (including the increasing problem of multiple antibiotic-resistant bacteria), such efforts seem to be prudent and long overdue.*

The prudently conservative tone of the *Wound Care* paper contrasts with Wolcott's enthusiasm when speaking only for himself. He believes the abundant evidence from eastern Europe shows that phage treatment "is not experimental," but effectively a proven success. "We worked with the FDA," he recalls. "We were trying to bring a group of phages for *staph* and *pseudomonas* to market, for Intralytix [a Baltimore company focused on phage products]. We tested 20 patients, with 20 more as a control, 40 patients in the study, and we had no adverse events. When I called the FDA, [they said] 'oh good, now we know that bacteriophages are safe.' And during this time frame, about 10 million doses of bacteriophages are given from pharmacies in eastern Europe, and you're saying that those 10 million cases, because they weren't in a randomized control trial, are meaningless? And your 20 patients in this tiny fledgling RCT is what tells western medicine that these are safe? It sounded so arrogant to me! One cohort study from Poland had 73,000 people from this town that had an outbreak of cholera. That's a big cohort."

He believes the regulators' attitude is partly "CYA (cover your ass). They can only get in trouble if they approve something and all of a sudden it causes harm to people."

While Wolcott hasn't had trouble bringing in enough phage to treat a few patients—most of them extreme cases for whom all other alternatives had failed—he complains that he can't obtain a sufficient supply to make phage treatment a regular or routine part of his practice. "I went to Tbilisi to a pharmacy and said I want to buy up all the phage you've got. It's very cheap. We just bought up all their bottles and brought them home. [But for large amounts] you need fermenters. You need a way to produce them. There's all these shipping rules. Like certain drugs we can get cheaper in Canada can't come down here. I can only bring it back for my personal use. The fermenters at Tbilisi turn out millions of doses a year, and you don't see a mass problem. There's no mutation. They're not growing horns or anything. It's bacteriophage. When you turn on your shower, two billion come over your body. They're growing on the biofilm in our bodies. They're our friends!

"And you don't think a year matters? But in a year several thousand people die ..."

3) ***Self treatment with phages imported from abroad.***

"People [in the U.S.] are allowed to bring in or have sent in such products for their own use for several months, using them with the help of their physician," said Evergreen College's Dr. Kutter, referred to earlier. "I think that is true throughout the country. I have also never had trouble bringing in phage with me through customs. In Washington (and I understand Oregon), Naturopathic physicians have the explicit right to use natural products that have been approved and extensively used elsewhere. That is a significant part of their training, on top of a lot of hardcore science and a variety of other modalities."

Canada and the U.S. have slightly different requirements for importation for personal use, but the Toronto woman mentioned above, in alternative (1), managed to import phages for her own use, over and above those she'd brought back from her hospital treatment in Tbilisi. She also sought support from a Naturopath, Dr. Jennifer Yun, of Toronto.

"As long as it's not on Schedule F [a list of banned substances], and you're an individual importing enough for your own use for a three-month period, that's it [it's permitted]," the woman said. At the time she went to Tblisi, there were too many different species of bacteria to be able to specify which phages were needed. It was only after being examined there, and her bacterial attackers identified, that she knew which phages to use. Once home, however, she could continue treating herself with phage mixtures prepared in Georgia.

"I used the Naturopath for the intravenous administration of vitamins, not to administer the phages," she added. "The hospital [in Canada] had missed that there was also anaemia going on, but the Naturopath picked it up."

"It's a bit of a grey area," said Dr. Yun. "It's not something I can prescribe directly, because there's no precedent set in Ontario, where my jurisdiction lies. Under the Drugless Practitioners Act my role with phage therapy is not that I would administer it. It's more that I can make a recommendation and the patient can go out and seek it on their own. They can order a personal supply. But because phages are considered to be medical interventions, we don't have access to it, as an MD would. My role is supportive, while they are undergoing the phage treatment, to support their body to assure that they have a decreased risk of reinfection."

~

Taken together, all three of the currently available treatment alternatives in North America, and much of the rest of the world, can at best account for a handful of patients—those lucky enough to hear of phage therapy, and courageous or desperate enough to try it.

Because phage therapy for human infections—as successfully developed in the former Soviet bloc, and tested in eastern Europe for decades—can't easily be patented by profit-oriented corporations in the West, or would cost them too much to obtain government approval from regulators who are either ignorant of the subject or afraid to permit something "new," it languishes. A possible tool to save thousands of people from antibiotic-resistant microbial infections, including lethal ones, goes largely unused.

Except to prevent contamination of a few food products, or to guard the health of shrimp.

Epic gouging

Nor is the antibiotic-versus-phage saga the only example of the pharmaceutical industry putting profit ahead of saving lives. One grotesque example of greed was furnished by Martin Shkreli, formerly with Turing Pharmaceuticals AG, when in 2015 he hiked the price of daraprim, a drug used to treat a parasitic condition known as toxoplasmosis, which attacks people with compromised immune systems, such as those with cancer or HIV. He abruptly raised the price of the drug by 5,000 per cent, to $750 per pill,[81] thus throwing the lives of patients who could not afford the price in danger. Toxoplasmosis can be "deadly for unborn babies and patients with compromised immune systems."[82]

The move caused widespread public outrage, but Shkreli later told interviewers "of course" he would do it again. "Everybody's doing it. In capitalism you try to get the highest price you can for a product."[83] Mocking those who found his move outrageous, he later auctioned off chances to punch or slap him in the face. He said bids went as high as $78,000.

His example appeared to be epic gouging, but going beyond what patients can bear is not all that unusual in the industry. It's pretty much mainstream. Mylan Pharmaceuticals, for example, hiked the price of the life-saving EpiPen, used for emergency treatment of persons undergoing severe allergic reactions, by 450 per cent since it bought the product from Merck in 2007.[84] The pens, which contain approximately $1 worth of epinephrine, are used by allergic persons who are reacting to such potentially fatal stimuli as eating peanuts or being stung by a bee.

According to reports, Mylan bought the EpiPen delivery system, priced originally at around $50, then "aggressively marketed the drug to concerned parents, while increasing prices annually" to around $600.[85]

In 2017, it was also revealed that U.S.-based pharmaceutical companies "had mounted a public relations blitz to tout new cures for the hepatitis C virus and persuade insurers, including government programs such as Medicare and Medicaid, to cover the costs. That isn't an easy sell, because the price of the treatments ranges from $40,000 to $94,000—or, because the treatments take three months, as much as $1,000 per day."[86]

These look like extreme examples, but a revealing article by Fran Quigley, director of the Health and Human Rights Clinic at Indiana University's Robert H. McKinney School of Law, makes the overall picture clear:

> *Pharmaceutical corporations consistently set most of their prices hundreds of times higher than their manufacturing costs, then relentlessly raise those prices at rates far exceeding inflation. The result is breathtaking corporate profits as high as 42 per cent annually. The industry's average return on assets more than doubles that of the rest of the Fortune 500.*
>
> *Meanwhile, many US seniors are forced to choose between paying for medicine or food. Hundreds of cancer patients recently took to the pages of the journal* Mayo Clinic Proceedings *to angrily protest the fact that with the cost of cancer medicines now averaging $100,000, one in five of their patients can't afford to fill their prescriptions.*[87]

The United States is by far the worst offender in this regard, with other developed nations, including Canada, offering drugs at much lower prices, and often subsidizing their purchase with government allowances. As Quigley notes: "US patients pay far and away the highest price globally for prescriptions—often twice as much as patients in other developed nations." As Don Reichmuth says: "This situation is really bringing out the greed."[88]

Nor is government, at least in the U.S., likely to do much to change things. The political muscle of Big Pharma, noted previously, assures this, and works to keep prices high. A recent bill in the U.S. Congress, the *21st Century Cures Act*, heavily supported by the industry, contains provisions to "extend patent life to delay the availability of lower priced generic drugs."

The battle for acceptance of bacteriophages is thus only part of a bigger picture, in which the greedy prey upon the sick or wounded, making money at their expense. Perhaps the extreme length to which this attitude can be carried, even more extreme than the tale of Martin Shkreli, is illustrated by two final examples: the marketing of OxyContin, and the dumping of banned drugs in the Third World.

Developed by a Canadian subsidiary of the U.S.-based Purdue Pharma, OxyContin was marketed as a super-efficient medication to treat pain "without unacceptable side effects." Only there were side effects, and at some point, as *Globe & Mail* writers Grant Robertson and Karen Howlett reported,[89] the company knew it—but kept mum. "Deadly problems were emerging with the drug, and the industry knew it." The report continues:

> *OxyContin wasn't merely a commercial success because it was effective at killing pain—it was also highly addictive. Patients who were prescribed the seemingly benign pills for everyday conditions, such as back pain, were becoming hopelessly dependent upon them, unable to break their habit and requiring stronger and stronger doses as time went on. Increasingly, people were dying.*
>
> *Canada's opioid epidemic, which traces its roots back to the introduction of OxyContin and Patent '738, has now killed thousands.*

The truth, eventually, did out:

> *The scope of the problem, and Purdue's deception, did not become publicly known until May 2007, when the company's U.S. parent and three of its top executives settled criminal and civil charges against them for misbranding OxyContin as less addictive than other narcotics. The company agreed to pay $634.5 million in fines—at the time, one of the largest settlements a pharmaceutical firm had paid for market misconduct.*
>
> *Purdue acknowledged in an agreed statement of facts that it "fraudulently and misleadingly marketed OxyContin as less addictive, less subject to abuse, and less likely to cause withdrawal symptoms than other pain medications."*[90]

Before the deception was exposed, Purdue had made more than $30 billion from sales of the drug.

As for dumping medications in Third World countries, after they've been banned in the developed world, Carolyn Nordstrom notes a typical example, which she encountered personally while travelling in southern Africa:

> *The malaria drug I'd been given (I later found out) had been banned by the WHO [World Health Organization] for causing heart damage and then had been dumped illegally—and without warnings—by large pharmaceutical companies in needy locales like Angola.*[91]
>
> *... As I walk through the vulnerable populations of war-torn countries, transitional nations, and economic empires, I observe that illicit pharmaceuticals constitute a domain as large as that of illegal narcotics in terms of profit.*[92]

In system terms

Finally, how does all of this medical history work itself out in terms of Systems Theory?

First, where phages and infections are concerned, several interlocking systems are involved, each affecting the others:

1) the human food production system, as organized under the factory farm model
2) the for-profit corporate pharmaceutical system
3) the human medical care system
4) the human body's metabolic system
5) the bacterial reproductive system
6) the overall microbial ecosystem in which phages are predators and bacteria their prey.

All these are in play simultaneously, along with a number of minor subsystems, too numerous to list here.

A general description of the overall scenario might go like this:

The factory farm system, whose product or output is protein in the form of fat cattle and pigs, as well as monetary profits for its owners, uses low doses of antibiotics as a growth stimulant for livestock. This creates two positive feedback loops, one generating more pounds of meat and the other more dollars. However, it also creates a third, unintended positive feedback loop, which could be seen as a subsystem in its own right, whose product or output is antibiotic-resistant bacteria.

The output of this third, unintended loop or subsystem becomes, in its turn, an unwelcome input for the human medical system, as well as for the human body's metabolic system, by pumping antibiotic resistant bacteria into them. The bacteria streaming in act as a negative feedback loop, or brake, for their target systems, rendering the output of the medical care system—protection of human health—more difficult, and slowing down or terminating the output of the human body's metabolic system by making people sick or killing them.

The corporate pharmaceutical system, which formerly dealt with microbial resistance by developing new antibiotics, for the most part no longer does so.

> **This system's main, or overriding output is actually the generation of money for its owners, and drugs are only a secondary output, used as vehicles to create the profits.**

Unfortunately, antibiotics have come to be seen as not sufficiently profitable, and so industry has largely ceased developing them. Phages, which cannot be patented, are also unlikely to generate super-profits, and in fact could become serious rivals of the increasingly ineffective antibiotics still in use. So they are kept off the market.

> **The profit motive thus acts as a negative feedback loop within the corporate system, slowing or halting development of drugs that aren't seen as sales blockbusters. Or, as seen in the case of OxyContin, marketing drugs that turn patients into addicts.**

Draper L. Kauffman, whose pioneering book, *Systems 1: an introduction to systems thinking*, has been cited several times in this chapter, puts it well: "You can never do just one thing ... in addition to the immediate effects of an action, there will always be other consequences of it which ripple through the system."[93] Or through neighbouring systems.

Unless some factor is introduced to control or halt such destructive inputs, they will simply continue and increase, causing general ruin and death, and leaving doctors, nurses and other medical personnel helpless. North American governments, more responsive to the corporate imperatives of Big Agriculture and Big Pharma than to the welfare of ordinary citizens, are reluctant to follow Europe's example and ban the use of antibiotics as growth stimulants, or readjust its regulatory criteria to allow phages.

Some hope may exist for the development of antibacterial phage spinoffs, in the form of chemical products derived from them, but these have yet to appear, could prove prohibitively expensive, and amount to finding a way to wield cat claws or teeth, rather than using cats. A clumsy solution.

The result, at least in the short run, is ***an efficient overall physical and socio/political system for the production of resistant bacteria, guaranteeing that microbes will continue to gain the upper hand, and people will become sick or die in ever larger and larger numbers.***

As for price gouging, and Big Pharma's political resistance to regulation, while introducing dangerous drugs such as OxyContin, these ***act as both a positive feedback loop, creating massive profits, and simultaneously as a negative loop, permanently curtailing the metabolic systems of thousands upon thousands of people.***

That is, killing them. One of the pharmaceutical system's chief outputs has become death.

Again putting profit above human health, the industry also makes it difficult, after original patents run out, for makers of cheaper, generic drugs to obtain samples needed to make their cheaper products. Major drug makers often withhold or refuse to provide such samples, thus delaying introduction of the generics and keeping prices artificially high. The FDA has listed "more than 50 drugs whose manufacturers have withheld or refused to sell samples, and cited 164 inquiries for help in obtaining them."[94]

In addition to all of these problems, recent reports have surfaced calling into question the safety of various medical devices, such as hip implants and coil birth control implants. "Hundreds of thousands of people in the world may have been exposed to toxic metals from 'metal on metal' hip implants," and others have complained of adverse effects from other such devices.[95] In the U.S., the FDA approval process for such devices is less stringent than the process for approving prescription drugs.

And the massive problem (at least for the United States) of medical insurance hasn't even been mentioned. As noted above in the discussion of Phagoburn, the only corporate industry that outdoes Big Pharma in its efforts to influence U.S. government policy is the insurance industry, which is among the chief reasons why Americans are nearly the only residents of an industrially developed country NOT to have comprehensive government provided medical care. Most European countries, as well as Canada, have had national health insurance for decades, and residents of those countries can only look on in astonishment as Americans struggle over the issue of so-called "Obamacare."

The version of government financed health care put in place by the former U.S. president's administration is nowhere near as comprehensive as its European or Canadian counterparts. But even this is struggling for survival as political conservatives, urged on by lobbyists, target it for repeal.

The profit motive cripples the U.S. system, and many other systems, at every turn.

Only a groundswell of public opinion can change things. "People will get it [phage therapy] only if they demand it," concludes Rosemonde Mandeville. And the same is true for fair drug prices, comprehensive government medical insurance, and a medical profession and medical schools that put health before money. They will come when people force the issue.

And not before.

If this were our only "resistance problem," it would be bad enough. But, as the following chapters will show, it's not the only one. There are other monsters out there, as big and bigger. And other systems generating them.

---

1 This popular children's rhyme, sung as part of a game in which the participants hold hands, then fall down, has often been depicted as originating with the Black Death outbreaks in Europe, and as describing what happens to plague victims. Folklorists appear to reject this theory, insisting that the now-widely-used version of the rhyme has a more recent origin. Whatever its source, the application is undeniably appropriate.

2 Schmid, Boris and Nils Christian Stenseth, "Plague outbreaks that ravaged Europe were driven by climate changes in Asia," The Conversation, 1 March 2015, as posted on Truthout, 1 March 2015, at: www.truth-out.org/news/item/29370-plague-outbreak-that-rava ... 1.

3 Andrew Nikiforuk, *The Fourth Horseman* (Toronto: Penguin Books Canada Ltd., 1991), 57.
4 Nikiforuk, *Op. Cit.*, 60.
5 *Loc. Cit.*.
6 Gordon L. Robertson, *Food Packaging: principles and practice*, (New York: Marcel Dekker, 1998), 187.
7 A very mild solution of phenol was, nevertheless, used to make a popular body soap, Lifebuoy, common in the U.S. as late as the 1950s. A similar soap was used in British schools as well, where it was employed as a punishment to "wash their mouths out" when students used bad language!
8 *Wikipedia, the Free Encyclopedia*, "Freischutz," as posted online 24 March 2014, at http://en.Wikipedia.org/wiki/Freischutz, 1. According to German folklore, a freeshooter (*freischutz*) made a contract with the devil for seven magic bullets, six of which would accurately hit any target, but the seventh of which was the devil's to use.
9 Guy E. Abraham, "The History of Iodine in Medicine, Part I: from Discovery to Essentiality," *The Original Internist*, spring 2006, 35.
10 *Wikipedia, the Free Encyclopedia*, "George H. Tichenor," as posted online 12 April 2014, at http://en.wikipedia.org/wiki/George_H_Tichenor, 1.
11 Draper L. Kauffman, Jr., *Systems 1: an introduction to systems theory*, (Minneapolis, MN: WS. A. Carlton, Publisher), 38-9.
12 Ernest J. Sternglass, *Low-Level Radiation*, (New York: Ballantine Books, Inc., 1972), 13-21. See also Ernest Sternglass, *Secret Fallout*, (New York: McGraw-Hill Book Company, 1981).
13 Thomas Pawlick, "The Silent Toll," *Harrowsmith Magazine*, June 1980, 33-49.
14 Thomas Hausler, *Viruses vs. Superbugs: a solution to the antibiotics crisis?*, (London: Macmillan, 2008), 64-5.
15 Kauffman, *Op. Cit.*, 38-9.
16 *Wikipedia, the Free Encyclopedia*, "Felix d'Herelle," as posted 12 August 2014, at http://en.wikipedia.org/wiki/Felix_d'Herelle, 3.
17 Thomas Hausler, *Op. Cit.*, 58-9.
18 *Ibid.*, 78.
19 *Ibid.*, 55.
20 *Ibid.*, 75-6.
21 *Ibid.*, 99.
22 *Wikipedia, the Free Encyclopedia*, "Penicillin," as posted 18 August 2014, at http://en.wikipedia.org/wiki/Penicillin, 7.
23 William Kalle Lasn, "The Uncooling of America: the history of corporations in the United States," excerpted from Culture Jam/Adbusters Magazine, 1999, as posted 25 August 2014 at http://thirdworldtraveler.com/Corporations/Hx_Corporations_US ..., 4.
24 *Wikipedia, the Free Encyclopedia*, "Pharmaceutical industry," as posted 13 May 2014 at http://en.wikipedia.org/wiki/Pharmaceutical_companies, 3.
25 Hausler, *Op. Cit.*, 136.
26 *Op. Cit.*, 155-6.
27 Anna Kuchment, *The Forgotten Cure: the past and future of phage therapy* (New York: Copernicus Books, 2012), 55.
28 Kuchment, *Op. Cit.*, 56.
29 *Op. Cit.*, 61.
30 *Wikipedia, the Free Encyclopedia*, "Phage Therapy," as posted 14 December 2012, at http://en.wikipedia.org/wiki/Phage_therapy , 2.
31 Cited by Kammerle Schneider and Laurie Garret, of the Council on Foreign Relations, in "Non-therapeutic use of antibiotics in animal agriculture, corresponding

resistance rates, and what can be done about it," *Center for Global Development, Updates*, 19 June 2009, p.1., as posted online 3 December 2014 at http://www.egdev.org /article/non-therapeutic-use-antibiotics-animal-ag … .

32 Kauffman, *Op. Cit.*, 38.

33 Robert Lawrence, "Antibiotic resistance: how industrial agriculture lies with statistics," *Huffington Post*, 1 October 2013, p.1, as posted online 3 December 2014, at http://www.huffingtonpost.com/robert-lawrence/antibiotic-resistance_ …

34 Hausler, *Op. Cit.*, 37..

35 Sabrina Tavernise, "Antibiotic-resistant infections lead to 23,000 deaths a year, CDC finds," *The New York Times*, 16 September 2013, as posted online 5 August 2015, at www.nytimes.com/2013/09/17/health/cdc-report-finds-23000-de …

36 The Alliance for the Prudent Use of Antibiotics, "The need to improve antibiotic use in food animals," March 2012, as posted online 3 December 2014 at http://www.tufts.edu/med/apua/about_issue/antibiotic_agri.shtml

37 Hausler, *Op. Cit.*, 42.

38 Will Hutton, "We risk disaster if drug giants don't invest in research," *The Observer*, 4 May 2014, p.2-5, as posted online at http://www.theguardian.com/commentisfree/2014/may/04/risk-disater …

39 *Australian Broadcasting Corporation*, "Antibiotics 1928-2000: before the miracle," 1999, as posted online at http://www.abc.net.au/science/slab/antibiotics/history.htm , 13 April 2014.

40 Projan, S.J. and Schlaes, D.M. "Antibacterial drug discovery: is it all downhill from here?" *Clinical Microbiology and Infection*, November 2004, Vol. 10, Issue Supplement s4, 18-22, as posted online 27 February 2014 at http://onlinelibrary.wiley.com/doi/10.1111/j.1465-0691.2004.1006.x/full

41 *Ibid.*

42 Erica Etelson, "Why there's no Ebola vaccine," *Truthout*, 12 December 2014, as posted at http://truth-out.org/opinion/item/28091-why-there-s-no-ebola-vaccine … 20 December 2014, 1.

43 *Op. Cit.*, 3.

44 Jennifer Yang, "Drug-resistant superbugs now a global threat, says WHO," *The Toronto Star*, 30 April 2014, as posted online at http://www.thestar.com/news/world/2014/04/30/drugresistant_bacteri …

45 "G. Eliava Institute of Bacteriophages, Microbiology and Virology, Tbilisi, Georgia," *Phage Therapy Center*, http://www.phagetherapycenter.com/pii/PatientServlet?command=stati …, as posted 14 December 2012.

46 Hausler, *Op. Cit.*, 175-6.

47 "Michel Chretien racantoe Felix d'Herelle," Commission de la capitale nationale du Quebec, les grands d'aujourdhui racontent ceux d'hier, 7 August 2008, as posted at www.newswire.ca/en/story/237727/dans-le-cadre-de-la-serie-le … 30 June 2015.

48 Timo Minassen, "The renewal of phage therapy to fight antimicrobial resistance–Part II: what about patent protection and alternative incentives?" *Bill of Health* blog, 7 August 2014, 1, as posted at http://blogs.law.harvard.edu/billofhealth/2014/08/07/the-revival-of-ph …

49 Karl Thiel, "Old dogma, new tricks–21st century phage therapy," *Nature Biotechnology*, vol. 22, 2004, 31-36, as posted online at http://www.nature.com/nbt/journal/cl22/n1/full/nbt0104-31.html

50 Thiel. *Op. Cit.*, 3.

51 Jonathan Caplin, "Bacteriophage therapy–old treatment, new focus?" *Microbiologist*, June 2000, 20-23.

52 52 Steven Projan, "Phage-inspired antibiotics?" *Nature Biotechnology*, volume 22, number 2, February, 2004, 168.

53 *Wikipedia*, Op. Cit., 6.
54 Richard Stone, "Stalin's forgotten cure," *Science*, volume 298, 25 October 2002, 6, as posted online 27 March 2013, at http://www.phageinternational.com/phagetherapy/ stalins_print.htm
55 *Loc. Cit.*
56 The editors, "PhageTherapy," *Microbe Wiki, the student-edited microbiology resource*, http://microbewiki.kenyon.edu/index.php/Phage_Therapy, 8, as posted online 28 December 2014.
57 Projan, *Op. Cit.*, 167.
58 Gary K. Schoolnik, William C. Summers and James D. Watson, "Phage offer a real alternative," *Nature Biotechnology*, vol. 22, number 5, May 2004, 506.
59 Hausler, *Op. Cit.*, 232.
60 *Op. Cit.*, 223.
61 Comments taken from a personal interview by the author, 1 August 2015.
62 According to the *Himmelfarb Health Sciences Library*, a randomized study or trial is "a study design that randomly assigns participants into an experimental group or control group. As the study is conducted, the only expected difference between the control and experimental groups … .is the outcome variable being studied." As posted at https://himmelfarb.gwu.edu/tutorials/studydesign101/rcs.html 12 August 2015.
63 See online summary: Andrzej Gorski et al, *A Historical Overview of the therapeutic use of bacteriophages*, Bacteriophage Laboratory & Therapy Center, L. Hirszfeld Institute of Immunology and Experimental Therapy, Polish Academy of Sciences, 53-114, Wroclaw, Poland, pp. 30-31, as posted 26 August 2015 at www.ema.europa.eu/docs/en_GB/document_library/Presentation/2015/06/WC500188391.pdf.
64 "Phagoburn: Evaluation of phage therapy for the treatment of burn wounds," Phagoburn website, as posted at http://phagoburn.eu/ 2 August 2015.
65 M. Asif Ismail, "Drug lobby second to none," Center for Public Integrity, 7 July 2005 (updated 19 May 2014), as posted at http://www.publicintegrity.org/2005/07/07/5786/drug-lobby-second-none, on 26 July 2015.
66 Matthew Harper, "The FDA is basically approving everything. Here's the data to prove it.," *Forbes*, 20 August 2015, as posted at http://www.forbes.com/sites/matthewharper/2015/08/20/the-fda-is-bas … 23 August 2015.
67 "Criticism of the Food and Drug Administration," *Wikipedia*, https://en.wikipedia.org/wiki/Criticism_of_the_Food_and_Drug_Admi … as posted 27 July 2015, 3.
68 Natalie Correia, Senior Marketing Coordinator, Ontario Blue Cross, personal e-mail, *Bacterial phages*, 21 August 2015, Natalie.Correia@ont.bluecross.ca
69 Maryann Schultz, Health Care Service Corporation, personal e-mail, *Media query*, 9 September 2015, Maryann_Schultz@bcbsil.com
70 Joanne Woodward Fraser, Senior Communications Advisor, Ontario Ministry of Health and Long-Term Care, personal e-mail, *Phage therapy*, 18 August 2015, Joanne.WoodwardFraser@ontario.ca
71 David Jensen, Media Relations Coordinator, Ontario Ministry of Health and Long-Term Care, personal e-mail, Bacteriophages, 17 June 2015, david.jensen@ontario.ca
72 This patient asked that her name not be used.
73 Molly Shen, *The Komo News*, "'You become desperate: obscure therapy saves woman's life.," 30 March 2013, as posted online at http://www.komonews.com/news/local/You-become-desperate-Unort … 22 April 2015.
74 *Loc. Cit.*
75 Emily Darby, M.D., personal e-mail, *Re: phage therapy in North America*, 16 May 2015, edarbymd@gmail.com
76 Molly Shen, *Loc. Cit.*

77 Emily Darby, *Loc. Cit.*
78 Elizabeth Svoboda, "The next phage," *Popular Science*, April 2009, 60.
79 *Loc. Cit.*
80 D.D. Rhoads, Wolcott, R.D., Kuskowski, M.A., Wolcott, B.M., Ward, L.S., Sulakvelidze, A., "Bacterophage therapy of venous leg ulcers in humans: results of a phase I safety trial," *Journal of Wound Care*, Vol. 18, No. 6, June 2009, 237-243.
81 *Reuters/the Guardian*, "Martin Shkreli announces turnaround on 5,000% price rise for drug," as posted online at www.theguardian.com/business/2015/sep/23/us-pharmaceutical- ..., 24 December 2016.
82 Christie Smythe and Keri Geiger, *Bloomberg*, "Shkreli, drug price gouger, denies fraud and posts bail," 17 December 2015, as posted online at www.bloomberg.com/features/2015-martin-shkreli-securities-f ... 24 December 2016, 3.
83 Jared S. Hopkins, "Martin Shkreli says 'of course' he'd raise drug price again," *Bloomberg,* 23 December 2016, as posted online at www.bloomberg.com/news/articles/2016-12-23/martin-shkreli ..., 24 December 2016, 1.
84 *The Daily Team, the Thom Hartmann Program,* "Big Pharma increased price of life-saving Epipen by over 450 per cent," 23 August 2016, as posted online at www.truth-out.org/opinion/item/37342-the-price-of-the-life-savi ..., 25 August 2016.
85 *Ibid.*
86 Annie Waldman, "Big pharma quietly enlists leading professors to justify $1,000-per-day drugs," 4 March 2014, *Consumer Reports*, as posted online at www.truth-out.org/news/item/39675-big-pharma-quietly-enlists- ..., 1.
87 Fran Quigley, *Truthout News Analysis*, "The $100,000-per-year pill: how US health agencies choose Pharma over patients," 5 August 2016, as posted online at www.truth-out.org/news/item/37111-the-100-000-per-year-pill- ... 5 August 2016, 1.
88 *Ibid.*, 3.
89 Grant Robertson and Karen Howlett, *The Globe & Mail*, "How a little-known patent sparked Canada's opioid crisis," 30 December 2016, as posted online at www.theglobeandmail.com/news/investigations/oxycontin/articl ..., 1 January 2017.
90 *Op. Cit.*, 3, 10.
91 Carolyn Nordstrom, *Global outlaws: crime, money and power in the contemporary world,* (Berkeley, Calif.: University of California Press, 2007), 131-2.
92 *Ibid.*, 137.
93 Kauffman, *Op. Cit.*, 38.
94 Sydney Lupkin, "Drugmakers blamed for blocking generics have jacked up prices and cost the U.S. billions," 29 May 2018, *Kaiser Health News/Truthout*, as posted online 30 May 2018 at www.truth-out.org/news/item/44616-drugmakers-blamed-for-bl ..., 1.
95 Amanda Holpuch, "The Bleeding Edge: behind the terrifying new Netflix documentary," 25 July 2018, *The Guardian*, as posted online 25 July 2018 at www.theguardian.com/film/2018/jul/25/the-bleeding-edge-netf ..., 2.

Chapter Two

FOOD IS FUNDAMENTAL

"Je vis de bonne soupe et non de beau langage
(It's good food and not fine words that keeps me alive)."
—Molière, *Les Femmes Savantes*

From the microbial world, where phages eat bacteria, all the way up the ladder of evolutionary complexity to our own rung, food is fundamental. Everything that lives eats. Or dies. And the agricultural systems that humans have developed across the millennia have enabled our kind not only to eat, but to create the complex social systems we lump under the general heading of "civilization."

Whatever threatens our food threatens not only our individual, physical lives, but those societies as well.

Such obvious facts shouldn't need repeating, yet in little more than half a century the so-called "factory food" regime, particularly in North America, appears to have separated humans from any real awareness of the source of their sustenance, while at the same time rendering the stuff it produces less and less nourishing, to the point where much of it is actually poisonous. Simultaneously, it is also destroying the soil, the air and the water, and disrupting most life forms in the regions where it dominates, provoking a host of resistance mechanisms. It will eventually destroy our very ability to raise nourishing food at all.[1]

The factory farm system didn't spring up overnight, however. It had roots extending back at least as far as the so-called "enclosure" movements of 16th century England, and likely even earlier, to a time when, in some men's minds, **the goal of farming—and farming systems—first ceased to be seen as producing food, and became instead to produce money.**

Hand to mouth

It wasn't always thus.

If the archaeologists and palaeontologists are right, our ancestors, once down from the trees, became hunter-gatherers, killing small or even mid-sized game for meat, and picking and eating a wide variety of wild plants, living literally from hand to mouth. At some point, they started what is now called "forest farming," weeding out competing plants that grew around their favourite wild goodies, and protecting these proto-gardens from other creatures that might eat them.

Eventually, some 12,000 or so years ago, they took the next step and started gathering and planting seeds, in ground they prepared especially to promote seed growth. They also adopted the practice of keeping the animals they had been spearing or trapping, like the aurochs (progenitor of modern cattle), in pens, raising them for both milk and meat. In short, they became farmers.

No longer nomads, following the game, they stayed in one spot, where their plants grew and their animal pens were set up. Instead of erecting makeshift, temporary shelters as they migrated, they could build permanent houses, first in family compounds and then in villages. Village life required rules, to keep the peace, and thus, little by little, year by year, the whole complexity of rules and customs that constitute human societies, including even architectural rules and customs, came to be.

For most of those 12,000 or so years, the basic unit of production was the family, or in some cases, particularly when grazing livestock were involved, the local village. And across most of the world, farmers employed what are today known as "organic" methods. These were arrived at by practical trial and error: what worked was adopted, what didn't wasn't. But they were far from primitive, at least in their later stages.

Expressed in system terms, the output of the methods that were developed, their whole point and purpose, was to produce food, for bodily survival. The positive feedback loops created to encourage that output included tillage, to provide a good seedbed, and the processes of weeding and cultivating,

> **to keep out competitor plants. The chief negative feedback loop, tending to reduce the system outputs, was gradual exhaustion of the biochemical soil nutrients needed by growing plants.**

A number of methods were employed to overcome this negative loop, one of the earliest and simplest being what is today called "slash-and-burn" farming. At one time, it was practiced throughout the tropics, from Africa to Vietnam, wherever climate, seasonal heavy rainfall and fast-growing wild vegetation helped it along. It is still practiced in some tropical or semi-tropical areas.

The *milperos* of Central and Latin America are typical slash-and-burn farmers. Despite its violent-sounding name (given to it by researchers from the northern, developed countries), their method makes sense within its traditional limits.

In the old days, before large-scale industrial lumbering and land-clearing had taken place, the farmers of a given village would clear just enough acres of nearby forest land to provide room for their subsistence crops. In the spring, at the tail-end of the dry season, they'd burn away the tree cover with controlled fires. Then they'd move in and cut down the remaining brush by hand, with their long-bladed machetes. The ashes and compost created by this process provided enough nutrients to the shallow tropical soil, enough fertility, to last for several seasons.

They then planted maize and beans—the maize taking nitrogen out of the soil and the beans putting it back in—as well as vegetables and melons. They scattered the dung from cattle and pigs on their gardens and fields. And for a few years they'd have good harvests. When the fertility of the soil gradually wore out, they'd move to a new location, clear a different patch of forest, and start over again.

Eventually, after many years, they would work their way through a region, finally ending up back at the spot where they first started. In the interim, this spot would have lain fallow, and eventually grown back, first with so-called "pioneer" species, then with a second growth of the original forest species the villagers had once burned away. Its soil nutrients restocked, it was ready for a new cycle of burning and crop planting.

> *potatoes, this crop may be substituted for a portion of the corn, thus increasing the cash crops at the expense of forage.*
>
> *Wheat generally proves a better crop in which to seed clover and the grasses than does oats. In most parts of this section of the country the grasses are seeded in the autumn and the clover seeded early in the spring. Further south, both clover and the grasses may be seeded in the autumn. The four staple crops above mentioned may be arranged into several rotations, with manure and fertilizers applied ...*[2]

A four-crop rotation was one of the simpler ones, with others employing six or more crops, all of which had to be timed carefully to fit the ever-changing vagaries of weather and markets.

Tillage and ploughing added another set of skills to the blend. One did not simply go over the ground willy-nilly with a sharp implement, then throw seed on it. There was a pattern to ploughing, that depended on the size of the field, depth of the topsoil and the crop to be planted. The ploughman, following behind a team of horses or oxen, always started with a backfurrow, down the centre of the field, to mark it off and make a straight line to follow. He'd then turn and come back in the opposite direction, laying the second backfurrow over against the first. The two furrows were made to stop several feet short of the field's end. Then the ploughman would go completely around the backfurrows, plough blades always facing inward towards them, toward the original line. This tended to push the soil in the field always toward the centre, producing a gentle "crown," that prevented puddling of rainwater and minimized erosion.

Very large fields would be broken up into "lands," of varying sizes, each with its own back furrow and crown. A variety of ploughing depths were used, depending on the soil type and the type of roots on the crop to be planted, to prevent soil compaction.

A so-called moldboard plough with anywhere from two to six blades was used for normal soil. For rocky land, whose rocks might break the shoe of a normal plough, a disc plough was used. The round blades of the disc plough would turn over soil, but ride up and over rocks. Ploughing was followed by disking and/or harrowing with various implements, to break the soil clods up into finer bits. After seeding, the field was gone over with a spike tooth harrow, to cover the seed.

Drainage ditches had to be dug all round the fields, or lands, connecting to watercourses. This required at least a working knowledge of hydrology. And of course keeping his horses healthy, or in more modern times his steam or diesel tractor running, called for yet another complex set of skills.

"Dumb farmers," indeed.

As for demographics, while precise figures don't always exist, it's safe to say that historically most humans have been farmers. In ancient Egypt, for example, "the vast majority of the population, probably more than nine tenths during the first two millennia of Egypt's history, lived on the land, in mostly self-sufficient village communities."[3] In the ancient Roman empire, far more people toiled on the land than in the cities, although in Italy itself most of the toilers were slaves, working on the large holdings, or *latifundia*, of the Patrician class. The late Roman Empire imported most of its food, particularly grain, from its colonies—including Egypt—where it was grown by the majority local populations.

In later times, with the notable exception of 17th century England (as will be shown shortly), the situation was pretty much the same. In France in the 1780s, for example, an estimated 85-90 per cent of the population was rural,[4] with most land owned by peasants. In the newly-independent United States of 1790, farmers made up roughly 90 per cent of the population, and more than half of Americans were still farming as late as the 1870s.[5]

Today, **less than two per cent of the American population farms,**[6] with roughly similar figures posted in Europe, including Britain, as well as Japan. So what happened to cause such an extreme shift? When and why did things change?

Improvement, for whom?

Historians seem generally to agree that the pivot-point came during the "enclosures" that began in England in the 1500s. At the time, land ownership in the British Isles was more concentrated than in most of

Europe, with members of an hereditary aristocracy "holding an unusually large proportion of land."[7] This land was worked not by individual peasant-proprietors protected by both custom and the legal fact of their ownership, but by tenants who paid rent. In addition, England had a better system of roads than most countries, which fostered development of regional and even national markets.

These factors encouraged English landowners to look for ways to make more money, by raising their tenants' rent and pushing them to sell more produce to the widening markets. As a result, in addition to the market for food, a market for land leases soon developed. "To meet economic rents in a situation where other potential tenants were competing for the same leases, tenants were compelled to produce cost-effectively, on penalty of dispossession."[8]

Landlords began referring to this process as "improvement," or "reform," which from a purely technological or economic standpoint, at least, was sometimes true. Tenant farmers became more efficient producers, or got booted out for non-payment of the now-higher rents. They adopted better methods, or developed better machinery with which to farm. They also took over from less efficient neighbours, consolidated the land, removing fences and wild lands in between, and planting only those crops that brought the highest prices. Enlarged, concentrated holdings increasingly became the norm. Some common lands, traditionally used by entire villages to grow hay or serve as pasture for their cattle, were also absorbed for private use.

Soon, however, this wasn't enough. **Improvement came to mean any change that would bring in more money for the landlord**, including a change in the uses to which land was put. In some cases, even the most efficient tenant farmer couldn't compete with those new uses.

Such was the situation with production of wool, for which there was a growing market. Ellen Meiksins Wood describes the process:

> *The first major wave of enclosure occurred in the 16th century, when larger landowners sought to drive commoners off lands that could be profitably put to use as pasture for increasingly lucrative sheep farming. Contemporary commentators held enclosure, more than*

any other single factor, responsible for the growing plague of vagabonds, those dispossessed, "masterless men" who wandered the countryside and threatened social order. The most famous of those commentators, Thomas More, though himself an encloser, described the practice as "sheep devouring men" ...

Enclosure surfaced as a major grievance in the English Civil War. In earlier phases, the practice was to some degree resisted by the monarchical state, if only because of the threat to public order. But once the landed classes had succeeded in shaping the state to their own changing requirements—a success consolidated in 1688, in the so-called Glorious Revolution—there was no further state interference, and a new kind of enclosure movement emerged in the 18th century, the so-called Parliamentary enclosures ... the extinction of troublesome property rights that interfered with some landlords' powers of accumulation took place by acts of Parliament.[9]

Wholesale enclosure became the rule, until by the 19th century "unenclosed commons had become largely restricted to rough pasture in mountainous areas and to relatively small parts of the lowlands."[10] And independent, small farmers soon became as rare in England as unenclosed common land.

Thus launched in Britain, the trend toward rural depopulation and the concentration of food production in fewer and fewer hands increased steadily around the world, spurred sometimes by political and sometimes by technological changes—changes which, at their base, often (though not always) had the accumulation of more money as a root cause.

An example of political change came barely a year after the shooting stopped in the American colonies' war for independence from Britain, when Shays' Rebellion broke out in Massachusetts. A postwar depression and credit squeeze, which led coastal merchants to foreclose on loans and the merchant-dominated state government to raise taxes, ended up hitting the subsistence farmers in the western part of the state hardest.

These farmers, many of them Revolutionary War veterans, "had little in the way of assets beyond their land, and bartered with one another for goods and services. In lean times, farmers might obtain goods on credit from suppliers in local market towns who would be paid when times were better."[11] The local suppliers were often partners with the coastal merchants and, when the squeeze came, passed on their demand for payment. "Individuals began to lose their land and other possessions when they were unable to fulfill their debt and tax obligations"[12] with hard currency. As times got tougher and the pressure got harsher, the farmers, inevitably, rebelled. They were put down by blunt, military force.

David Szatmary, whose book *Shays' Rebellion* is considered by many the definitive work on the subject, concludes:

> *The rebellion represented a dynamic process that had its basis in the continuing struggle between a largely subsistence, family-based, community-oriented culture of independent farmers and an acquisitive, individualistic way of life dominated by merchants, professionals, speculators and commercial farmers. In the larger context of American social development, the insurrection illustrated the tumultuous effects of the transition from traditional society to merchant capitalism.*[13]

Examples of scientific and technological changes, also pushed, at least in part, by the desire to make more money, are even more abundant.

Mechanical and Green Revolutions

England again proved an international trendsetter in what has been dubbed the British Agricultural Revolution, a subset of the Industrial Revolution. Catalyzed by the Enclosure Movements, which peaked in the early 1800s, it saw an unprecedented stream of inventions and discoveries almost too numerous to list. A sample:

> *Machines were invented to improve the efficiency of various agricultural operations, such as Jethro Tull's seed drill of 1701 that mechanized seeding at the correct depth and spacing, and Andrew Meikle's threshing machine of 1784. Ploughs were steadily improved,*

from Joseph Foljambe's Rotherham iron plough in 1730 to James Small's improved 'Scots Plough' metal in 1763. In 1789 Ransomes, Sims & Jefferies was producing 86 plough models for different soils. Traction machines also began to replace horsepower on the farms in the 19th century.

The scientific investigation of fertilization began at the Rothamsted Experimental Station in 1843 by John Bennet Lawes. He investigated the impact of inorganic and organic fertilizers on crop yield, and founded one of the first artificial fertilizer manufacturing factories in 1842. Fertilizer, in the shape of sodium nitrate deposits in Chile, was imported to Britain by John Thomas North as well as guano (bird droppings). The first commercial process for fertilizer production was the obtaining of phosphate from the dissolution of coprolites in sulphuric acid.[14]

And, of course, "the first commercially successful gasoline powered, general purpose tractor," built by Daniel Albone, appeared in 1901.[15]

It should be noted here that most of these advances were just that, namely advances in human knowledge and innovation. Only a Luddite might bemoan them. But it should be recalled that even the original Luddites (supposedly named after Ned Ludd, a textile worker who smashed two stocking frames in 1779),[16] from whom the pejorative term for "opponent of new technologies or technological change"[17] got its name, were not against machinery *per se*. Skilled weavers and other textile artisans who were threatened with unemployment by the introduction of new machines in English factories, they went about breaking looms and lacemaking machines, not in a vain effort to turn back the clock, but to gain "a better bargaining position with their employers"[18] and negotiate some sort of transitional settlement.

The new techniques and new technologies of the Industrial Revolution were being introduced willy-nilly, with little or no consideration for their possible social effects, and no overall planning for the welfare of those who would be affected. In too many cases, working people were simply thrown out of their jobs and subsequently their lodgings, into the street, with nowhere to go and no means of support.

In other words, **the developing industrial system, which included "efficient" farming, had created a new, positive feedback loop, whose output was jobless, homeless people.** Streaming out from the countryside, they converged on major urban centers like London, in hopes of finding work there, leading to the creation of the hideous, overcrowded slums so graphically described in the novels of Charles Dickens. It would be decades before the urban-based manufacturing system was able to absorb these migrants, and even when it did so, they were often far worse off than they had been on the land.[19]

> **The violent reaction of the Luddites, and other groups affected by mechanization, was a failed attempt to create a negative feedback loop, to slow down the system's output and give society time to find ways to absorb the surplus labour. The industrial system, which now included industrial-scale farming, had obeyed Newton's Third Law, and created its own human resistance.**

For more than 100 years, invention followed invention, enabling fewer and fewer farmers to work ever larger and larger farms, and forcing yet more people to migrate to the cities. Then, in the mid-20th century, came the mother of all advances, the so-called Green Revolution.

American plant geneticist Norman Borlaug is generally credited with the revolution's launch, through his efforts in crossbreeding new wheat and maize varieties while working in Mexico from the early 1940s onward, first with the Cooperative Wheat Research and Production Program, and later with the Consultative Group on International Agricultural Research (CGIAR). Borlaug and his teams produced thousands of new, high-yielding and disease-resistant varieties, which, when combined with large scale, mechanized methods, and the application of concentrated, inorganic fertilizers, increased crop yields per acre by phenomenal amounts.

Others followed Borlaug's lead, developing hybrid varieties of dozens of other crops, permitting bigger yields across the board. In the northern, industrially developed countries, this led to still greater out-migration from already depopulated rural areas, and to the expansion of farms to hitherto undreamed-of sizes.

Monoculture, the devotion of huge swaths of land to a single, high-yielding, price-profitable crop, planted year after year in the same place without rotation, became the rule. The inevitable depletion of soil nutrients caused by the lack of rotations was offset by continuing applications of highly concentrated inorganic fertilizers. As for the unnaturally plentiful banquet afforded to crop pests by producing their favourite foods in such quantity, this was offset by heavy use of inorganic chemical pesticides. The same strategy was used for weeds, which were eradicated by the application of increasingly lethal inorganic herbicides.

Agribusiness

In the short run, all of this looked beneficial, and to some degree it obviously was. A very few farm workers could produce enough food for a very large number of people. In the northern countries, where urban manufacturing and service industries had grown large enough to absorb what was by now a mere trickle of surplus farm workers, food was both plentiful and—distributed by mass market "super" chainstores—comparatively cheap. Companies engaged in growing and distributing food made millions, creating the economic powerhouse eventually dubbed "Agribusiness."

Particular regions specialized in certain crops for which their soils or climate were ideally suited, and steadily expanding acreages of these crops, from strawberries in California or corn in Iowa, to sugar cane, coffee or tea in the tropics, were planted in money-spinning monocultures around the world. The assembly line methods pioneered in the manufacturing sector by automakers (and tractor makers) like Henry Ford were adopted on factory livestock farms. Cavernous barns were filled with millions of so-called "battery" chickens, cramped in tiny cages where they could barely move, pecking at feed rations running past them on conveyor belts (see below), while equally cramped hogs, or "finishing" beef cattle were jammed together in gigantic outdoor feedlots, living in their own shit until shipped to slaughter.

Just as iodine, then antibiotics had become the Magic Bullets of the medical care system, so high-intensity agriculture captured the imaginations of industry planners everywhere, while the campaign fund contributions of the corporations engaged in it soon captured the loyalty of politicians as well. As with pharmaceuticals, where mergers and cutthroat competition gradually whittled down the number of players, so the number of agribusiness corporations gradually shrank, until the industry was dominated by a handful of majors.

The history of "improvement" in the northern countries was eventually repeated in the global South, where local subsistence farms, which had once produced enough food to feed their local populations, were quickly displaced in the rush for cash. Non-food export crops, following a pattern set by French rubber planters in Indochina decades earlier, often replaced food as the products of these newer, bigger operations. Instead of maize and beans for local villagers, the landscape was covered by sisal, sugar cane, hemp, tea and coffee plantations, whose products flowed north, to Europe, Japan and North America.

This newer trade pattern, which grew quickly in the years following the Second World War, got a massive booster shot in 1994 with the signing of the General Agreement on Tariffs and Trade (GATT), and shortly after by the establishment of the World Trade Organization (WTO). The terms of trade that were established, and which continue evolving in a similar direction today, tended strongly to favour the wealthy, manufacturing countries of the global North, providing them with relatively cheap raw materials or with tropical foods, like bananas or pineapples, that they couldn't grow at home, and removing any local regulations that might protect the small farmers of the global South.

GATT critics Tim Lang and Colin Hines summed up what was actually happening:

> *The model implicit in the GATT's agricultural deal is the Northern model of industrialized efficiency. Self-reliance is out, trade is in. National barriers to protect local production are to be dismantled ... If the North's development model is imposed and fosters the same rural/urban spread of population already seen in some of the richest,*

"efficient" agricultural systems like Canada, Australia or the U.K., 1.9 billion of the planet's rural dwellers will end up living in towns. To do what? Fed by whom?[20]

They concluded: "The large farmer, traders and big companies benefit, but the overall evidence is that intensive, high-input farming, the logical outcome of these policies, is disastrous for the environment, rural economies, food quality and food security."[21]

And that's not all. If Lang and Hines were trying to say the figurative agri-food emperor has no clothes, they only got as far as his hat and shirt. Far worse has been revealed over the years since.

Rather than repeat in complete detail what has been explained elsewhere,[22] a summary should suffice:

1. ***Socio/economic effects***

Unemployment/mass migration: As already noted, the Factory Farm System, or more accurately cluster of systems, produces an initial stream of rural refugees, who leave the countryside and flock to urban areas in search of work. If insufficient provision is made to absorb them, a short-term crisis occurs, **taking the form of positive feedback loops that produce unemployment, mass migrations, slums and spikes in the crime rates.**

A modern example of this unemployment loop occurred in the 1960s, when low-wage, landless Mexican migrant workers, who had been actively recruited by fruit and vegetable growers in California and Florida in the 1950s, were confronted with the invention of the mechanical tomato harvester. The harvester, developed in 1962 partly in response to demands by the United Farm Workers union for higher wages, brought quick results: 1,153 machines replaced 50,000 field workers. Ironically, the expensive machines, costing more than $80,000 each, also devastated growers' ranks. In 1962, there were 6,000 tomato farmers in California, but nine years later only 600 were able to afford to stay in business.[23]

In the industrial north, **the negative feedback loop of urban manufacturing has historically eased such problems, by eventually**

providing city jobs for both migrant fruit pickers and jobless ex-farmers. Today, however, many of those American or European urban manufacturing jobs have been outsourced to China, while in the Global South the international terms of trade make the growth of any indigenous manufacturing difficult. The result is a restless, impoverished and politically unstable mass of potential criminals or revolutionaries, piling up in Third World cities, or joining a mass out-migration.

These migrant populations are creating blowback for the industrial north, in the form of illegal Mexican or Central American migrants (no longer recruited by growers, and no longer absorbable by vanishing urban manufacturers) crossing into the southern U.S. states, as well as Africans crossing the Mediterranean to Italy and southern Europe. Right-wing U.S. politicians like the U.S. president Donald Trump are already demanding ever-harsher methods to keep the wetbacks out, from creation of a veritable Berlin Wall stretching across the U.S./Mexican border, complete with electric fences and buzzing, unmanned drones overhead, to shoot-to-kill patrols of local vigilantes. And every day, boatloads of would-be African or Arab migrants are found, drifting and abandoned or sinking in the Mediterranean waters off southern Europe.

The continuing out-migration of experienced, traditional family farmers from rural areas of Europe and North America, and of farm workers from the global South, poses a stark question for society: **If this constant loss of expertise continues, who will be left who knows how to farm with traditional, organic methods?**

Third World peasants may not be familiar with the latest science, but they at least know how to rotate a crop. As for North American or European farmers, who have access to the latest methods, particularly organic ones, they are becoming thin on the ground., as *Yes* magazine reports:

> *In 2012, the average age of American farmers was 58, according to the U.S. Census of Agriculture. In the same census, one-third of farmers were age 65 and older; only six per cent of farmers were younger than 35.*

And fewer new farmers are staying with it. In 2012, not quite 470,000 farmers had been on their land less than a decade—a 19 per cent drop from the number of new farmers just five years before.[24]

2. ***Inhumane treatment of animals***

The factory livestock operations mentioned above have led to grossly inhumane treatment of the living creatures passing through them. Karen Davis, of United Poultry Concerns, Inc., has described the plight of battery hens in detail,[25] only a small portion of which follows:

> *Battery hens live [for 10-12 months] in a poisoned atmosphere. Toxic ammonia rises from the decomposing uric acid in the manure pits beneath the cages, to cause ammonia-burned eyes and chronic respiratory disease in millions of hens ... Hens to be used for another laying period are force molted to reduce the accumulated fat in the reproductive systems by forcing the hens to stop laying for a couple of months. In the force molt, producers starve the hens for four to 14 days causing them to lose 25 to 30 per cent of their body weight along with their feathers. Water deprivation, drugs such as chlormadinone, and harsh light and blackout schedules can be part of this brutal treatment.*
>
> *Even eating is gruesome for the battery hen, who must stretch her neck across a 'feeder fence' to reach the monotonous mash in the trough, a repeated action that over time wears away her neck feathers and causes throat blisters. In addition, the fine mash particles stick to the inside of the hen's mouth, attracting bacteria causing painful mouth ulcers. A mold toxin, T-2, can taint the mash creating even more mouth ulcers in the hens, who have no choice but to consume what is in front of them.*
>
> *Battery hens are debeaked with a hot machine blade once and often twice during their lives, typically at one day old and again at seven weeks old, because a young beak will often grow back. Debeaking causes severe, chronic pain and suffering researchers compare*

> *to human phantom limb and stump pain. Between the horn and bone of the beak is a thick layer of highly sensitive tissue. The hot blade cuts through this sensitive tissue, impairing the hen's ability to eat, drink, wipe her beak, and preen normally ...*
>
> *At the end of the laying period, the hens are flung from the battery to the transport cages by their wings, legs, head, feet, or whatever is grabbed. Many bones are broken. Chicken 'stuffers' are paid for speed, not gentleness. Half-naked from feather loss and terrified by a lifetime of abuse, hens in transit embody a state of fear so severe that many are paralyzed by the time they reach the slaughterhouse. At slaughter, the hens are a mass of broken bones, oozing abscesses, bright red bruises, and internal haemorrhaging making them fit only for shredding into products that hide the true state of their flesh and their lives, such as chicken soup and pies ...*

And so on. The complete description goes on for pages, and reads like a true-life horror story. The lives of equally tortured hogs, or of "finishing" beef cattle jammed together in gigantic outdoor feedlots, or CAFOs (Confined Animal Feeding Operations), living in their own shit until shipped to slaughter, are no better, nor are the brutal, inhumane ways in which they are at last put out of their pain.

3. ***Food quality/nutritional effects***

The food products distributed by the Factory Farm system have over the years become steadily:

a) *Less varied*—Supermarket shelves contain fewer and fewer individual varieties of any given foodstuff, from apples to tomatoes. Where local farmers' markets and independent grocers once stocked dozens of varieties, grown locally on truck farms, chain stores now typically display only a handful. These are chosen to meet the convenience of industrial growers operating on a mass scale.

For example, to prevent damage to fruit shipped across the continent in trucks, growers select only tomato varieties with thick pericarp (outer layer) walls, that can withstand jostling without bruising. Thick-walled, rubbery tomatoes also proved desirable in eliminating labour costs for the *braceros* who used to hand pick fruit in the fields, mentioned above under socio/economic effects. Only rubber tomatoes could survive machine picking.

Other qualities deemed desirable in fruits and vegetables include high yield (in pounds per acre), uniformity of ripening time, and uniformity of size and colour, to provide a pleasing cosmetic appearance on supermarket shelves. Tomatoes, apples, peaches or other items that don't score at the top in terms of these characteristics have been largely discarded. Only flavour and nutrition are normally left off the list of sought-for traits.

The result of these imperatives? Of the thousands of existing varieties of tomato (a conservative estimate might put them at approximately 6,000[26]), supermarket chains typically offer only three or four. In all of North America, "more than 85 per cent of the tomatoes shipped for processing into canned or other products come from California," where only 10 varieties "accounted for more than 60 per cent of the entire processing tomato market. Five of these were proprietary varieties, developed by major multinational food processing companies that require their contract suppliers to grow only their in-house varieties."[27] As for fresh tomatoes, 50 per cent of which come from Florida, "11 varieties dominated the fresh market, with only five accounting for more than 80 per cent of all Florida tomatoes grown. In the EU, which exports to Canada in winter, the story is much the same. In Portugal in 1999, for example, more than 80 per cent of the tomato crop was accounted for by only six varieties."[28]

At the turn of the last century, "there were more than 7,000 apple varieties grown in the United States. By the dawn of the 21st century, over 85 per cent of these varieties, more than 6,000, had become extinct."[29] By the year 2000, 73 per cent of all lettuce grown in the U.S. was one variety, iceberg.[30]

A study conducted by the Rural Advancement Foundation International "compared USDA listings of seed varieties sold by commercial seed houses in 1903 with those in the U.S. National Seed Storage

Laboratory in 1983. The survey, which included 66 crops, found that about 93 per cent of the varieties had gone extinct."[31]

The same is true for most fruits and vegetables, as well as for the diminishing number of livestock breeds that supply our meat, milk and eggs.

And the advent of Genetically Modified (GM) plant varieties will cut down still further on diversity, particularly where crops that require open pollination are concerned. Once a GM variety, for example of apple or alfalfa, is planted in a region, it will be impossible to prevent bees or other insect pollinators, or the wind, from carrying pollen from the GM plants to neighbouring fields, contaminating any non-GM varieties grown there.

Organic growers will be especially vulnerable, as they will be unable to guarantee that their crops are the pure variety they originally planted. Eventually, GM varieties will take over entire regions. Some see a deliberate policy in the process: "Contamination is an intentional strategy," said the University of Washington's Dr. Philip Bereano. "It's an intentional strategy by both the government and the industry. We have statements to that effect. Contamination in the field by pollen flow; contamination in the processing. They use the same railcars for engineered and non-engineered crops."[32]

The same article that quoted Bereano continued: "Ronnie Cummins, with the Organic Consumers Association, also discussed this in a recent interview, warning that any alfalfa growing within a five-mile radius of GM alfalfa will immediately become contaminated. The ramifications of this contamination are actually far worse than you might think, because alfalfa is a major food source for organic dairy cows. So once organic alfalfa becomes contaminated, organic milk and beef goes out the window too. Echoing Dr. Bereano's beliefs exactly, Cummins also said:

> *I believe this is an act of premeditated genetic pollution of the gene pool of alfalfa and related plants ... they understand that if you pollute enough alfalfa across the country to where it becomes impossible to grow organic alfalfa that isn't contaminated, perhaps then the organic community will weaken and allow genetically engineered animal feed under the rules of organic production.*

"Genetically modified canola (rapeseed) has escaped from the farm and is thriving in the wild across North Dakota, according to a study that indicates there are plenty of novel man-made genes crossing the Canada/U.S. border," the *Saskatoon Star Phoenix* reported. "GM canola was found growing everywhere from ditches to parking lots, the scientists report, with some of the highest densities along a trucking route into Canada ... research also found 'escaped' GM canola is becoming common on the Canadian prairies."[33]

> **In system terms, the modern, factory farm model, whose goal is to produce money, acts as a negative feedback loop within our overall food production system, restricting its output in terms of variety. "Restrict" is probably too mild a term. This negative loop acts to destroy variety outright.**

In future, if global climate disruption creates conditions hostile to some economically important crop varieties, the existence of other varieties which are more drought, temperature or moisture tolerant, could be crucial. If they have disappeared in the interim, where will we turn?

b) ***Less nutritious***—Every few years, in Britain, Canada and the U.S., sets of tables are published showing the nutritional content of foods, based on extensive testing in government laboratories. Researchers collect samples of foods from their countries' supermarket shelves, then subject them to close chemical scrutiny, to obtain the published figures.

A comparison of the tables published over the years (in 1950, 1963, 1975 and 2002), in all three countries, shows a steady—in some cases precipitous—decline in most basic nutrients, along with a simultaneous increase in substances generally considered problematic for human health.

For example, tables published by the United States Department of Agriculture (USDA) in 2002[34] show that "Skinless, roasted white chicken meat has lost 51.6 per cent of its vitamin A since 1963. Dark meat has lost 52 per cent. White meat has also lost 19.9 per cent of its potassium, while dark meat has lost 25.2 per cent. And what has chicken gained?

Light meat, 32.6 per cent fat, and 20.3 per cent sodium; dark meat, 54.4 per cent fat and 8.1 per cent sodium."[35]

Dairy products, favoured by consumers precisely because they are seen as a low-fat source of calcium and phosphorus to maintain strong bones and teeth, "gained 7.3 per cent fat since 1963, while losing 36.1 per cent of its calcium, 13.1 per cent of its phosphorus and–incidentally–fully 53.3 per cent of its iron. And what has it gained, besides fat? 76.85 per cent sodium."[36]

[A move is even afoot to create genetically engineered (GE) dairy products, such as milk, by altering the DNA of yeast or algae, then putting them in vats of sugar or sugar substitute to grow. A milk substitute called Muufri has been created in the laboratory,[37] using six key proteins of milk, but not including its other ingredients. At this writing, however, no one has examined what happens when various elements of a food substance are missing. "We don't always know which components of food have inherent health value or work together with other components to magnify effects," cautions Kerstin Lindgren of the Organic Consumers Association.]

The food tables also show that the Canadian potato has lost 57 per cent of its vitamin C in the past 50 years, while its American counterpart has lost 16.9 per cent just since 1963. Broccoli has lost 45 per cent of this crucial nutrient since 1963.

Vitamin C is needed by the human body to help control the development of cancer-causing "free radicals," and has been shown in the past to prevent such serious conditions as scurvy, once a scourge for sailors on long voyages where their rations lacked the vitamin. Iron, in such sharp decline in milk, prevents anaemia.

According to the Toronto *Globe & Mail*, the Canadian potato has also "lost 100 per cent of its vitamin A, which is important for good eyesight; 57 per cent of its iron, a key component of healthy blood, and 28 per cent of its calcium, essential for building healthy bones and teeth. It has also lost 50 per cent of its riboflavin and 18 per cent of its thiamine ..."[38]

These were the figures as of 2002. Since then the story has only gotten worse. All of the other declining nutrients are also needed, either to maintain normal body functions, or to ward off diseases. Yet virtually all continue to decline, across the board, in country after country where intensive agriculture is practiced.

c) ***More contaminated***–They say nature abhors a vacuum, and it appears that as fast as nutrients are disappearing from modern, mass-market foods, they are being replaced by a witches' brew of additives, pollutants, adulterants and poisons so long it would take a dictionary or encyclopaedia to list and describe them.

In fact, several such books have been written, including two personal favourites: *A Consumer's Dictionary of Food Additives*, by Ruth Winter (New York: Three Rivers Press, 2009), and *Foods That Harm, Foods That Heal: an A-Z guide to safe and healthy eating* (Montreal: Reader's Digest Association (Canada) Ltd., 1997).

The Center for Science in the Public Interest (CSPI) has also posted a list of common additives, rating them for safety, on its Chemical Cuisine website, at www.cspinet.org/reports/chem-cuisine.htm

A mere sampling of some of the more ubiquitous substances might include:

Acrylamide–a derivative of acrylic acid, used industrially to make adhesives and textiles, and identified by the World Health Organization (WHO) as a probable cancer-causing agent in laboratory animals. Apparently produced as part of manufacturing processes that require heating to make potato chips, french fries, crackers and other processed foods, it supposedly is not found in levels toxic to humans. However, at the time of its discovery in foods in 2002, it was not known what levels might be dangerous.

Antibiotics–Already discussed in Chapter One, these can be harmful both by provoking resistance in disease-causing bacteria they encounter after ingestion, and by destroying beneficial bacteria in the human digestive tract.

Dioxin (TCDD)–Made infamous as a component of Agent Orange, the herbicide widely sprayed over Vietnam during the American phase of the Indochina wars, and by its effects on local populations, it is described as "a toxic, cancer-causing chemical. Initial exposure to this agent can produce chloracne [generalized acne], liver injury and peripheral neuropathy [muscle weakness, impaired weakness and numbness, stinging and burning sensations]."[39] It has been branded (though its manufacturers dispute this) as a cause of severe birth defects.

Genetically Modified Organisms (GMOs), or material—Foods created, or altered by genetic manipulation are becoming increasingly common, particularly in North America where regulations governing their introduction and labelling are less strict than in Europe. As noted above, GM plant varieties that require open pollination, once planted will be impossible to keep out of neighbouring fields. This means that *even where they are not planted by a farmer they may, due to insect or wind contamination of the intended crop, still end up on the farm customers' plates.*

An advertisement published as far back as 1999 by the Turning Point Project, titled "Unlabeled, untested ... and you're eating it," summed up the situation then: "while there have been no tests so far conclusively establishing that genetically engineered foods are harmful to humans, the potential dangers are significant enough to mandate long-term independent testing of GE food products before release into supermarkets."[40]

The words "conclusively establishing" carried an eerie echo of the earlier tobacco wars, when cigarette industry apologists constantly repeated the mantra that there was no "conclusive link" between smoking and cancer, or smoking and heart disease. Perhaps no one will ever be able to prove that "smoking his 12,401st cigarette definitely caused Joe Smith's fatal lung cancer," but the weight of evidence eventually became overwhelming. And so it may well be with GMOs.

The Turning Point ad cited what was already known at the time:

TOXICITY. According to some FDA scientists, the genetic engineering of food may bring "some undesirable effects such as increased levels of known naturally occurring toxicants, appearance of new, not previously identified toxicants, increased capability of concentrating

> *toxic substances from the environment (e.g. pesticides or heavy metals), and undesirable alterations in the levels of nutrients." In other words, scientists from the FDA itself suspect that genetic engineering could make foods toxic.*
> *ALLERGENIC REACTION. FDA scientists also warn that genetically engineered foods could "produce a new protein allergen" or "enhance synthesis of existing plant food allergens." And a recent study in the* New England Journal of Medicine *showed that when a gene from a Brazil nut was engineered into soybeans, people allergic to nuts had serious reactions. Without labelling, people with certain food allergies will not be able to know if they might be harmed by the food they're eating.*
> *ANTIBIOTIC RESISTANCE. Many GE foods are modified with antibiotic resistant genes; people who eat them may become more susceptible to bacterial infections. Commenting on this problem, the British Medical Association said that antibiotic resistance is "one of the major public health threats that will be faced in the 21st century."*
> *CANCER. European scientists have also found that dairy products from animals treated with bovine growth hormone (rBGH) contain an insulin-like growth factor that may increase the risk of breast cancer, as well as prostate and colon cancer.*
> *IMMUNO-SUPPRESSION. Twenty-two leading scientists recently declared that animal test results linking genetically engineered foods to immuno-suppression are valid.*[41]

Studies in the 20 years since that ad appeared have continued to build the case, just as similar reports did in the old, tobacco wars days. Some examples:

> *The first ever GM food safety study to test over the entire life span of laboratory rats (2 years) found serious health impacts from eating Monsanto's genetically engineered corn NK603, which was approved in Canada in 2001. The peer-reviewed study also tested the impacts of consuming residues of Monsanto's herbicide Roundup, the widest selling herbicide in the world ...*

> *The study, published in the scientific journal* Food and Chemical Toxicology, *is the first animal feeding trial conducted over the lifetime of rats (700 days). Health Canada evaluates the safety of GM foods based on industry studies, the longest of which have been 90-day animal feeding trials ... Monsanto's GM corn NX603 is herbicide tolerant, meaning it is genetically engineered to withstand sprayings of Monsanto's herbicide Roundup ...*
>
> *The new study observed that rats fed the GM corn, or Roundup, developed tumours faster and died earlier than rats fed non-GM corn. The first tumour was observed after 120 days, with the majority detected after 18 months. The study shows GM corn can cause severe negative health effects in laboratory rats including mammary tumours and kidney and liver damage, leading to premature death.*
>
> ** Fed GM corn or Roundup, up to 50 per cent of males and 70 per cent of females died prematurely, compared with only 30 and 20 per cent in the control group.*
>
> ** Females developed fatal mammary tumours and pituitary disorders. Males suffered liver damage, developed kidney and skin tumours and problems with their digestive systems.*
>
> ** Rats fed GM corn and Roundup developed 2-3 times more tumours.*
>
> ** By the 24th month, 50 to 80 per cent of the females had developed large tumours compared to 30 per cent in the control group.*[42]

Far more disturbing in its implications was the discovery by Purdue University plant pathologist Dr. Don Huber of a microscopic pathogen **entirely new to science**, found in high concentrations of so-called Roundup Ready (namely, genetically engineered to tolerate the glyphosate herbicide Roundup) corn and soy. Shocked by his discovery, Huber wrote a letter to the then US Department of Agriculture's Secretary, Tom Vilsack:

> *For the past 40 years, I have been a scientist in the professional and military agencies that evaluate and prepare for natural and*

> *manmade biological threats, including germ warfare and disease outbreaks. Based on this experience, I believe the threat we are facing from this pathogen is unique and of a high risk status. In layman's terms, it should be treated as an emergency.*[43]

According to news reports, Huber "called for an immediate moratorium on approvals of all Roundup Ready crops," but barely two weeks after his letter was received, the USDA fully deregulated Roundup Ready alfalfa, and shortly afterward Roundup Ready sugar beets.

> *The pathogen is about the size of a virus and reproduces like a microfungal organism. According to Huber, the organism may be the first microfungus of its kind ever discovered, and there is evidence that the infectious pathogen causes diseases in both plants and animals, which is very rare.*
>
> *Laboratory tests show that the pathogen is present in a 'wide variety' of livestock suffering from infertility and spontaneous abortions. Huber warned that the pathogen could be responsible for reports of increased infertility rates in dairy cows and rates of spontaneous abortions in cattle as high as 45 per cent.*[44]

There are many more such reports and studies, but citing more would only belabour the obvious. The infant science of gene manipulation may be promising and hold many good things in future, but it unquestionably also holds dangers so great **that leaving its development in the hands of corporations intent only on making money, and with inadequate oversight by regulators, is folly.**

Nitrate/nitrite—Nitrates change into nitrites on exposure to air. Used to stabilize colour in curing meats, and also present in nitrate fertilizers, they can combine with stomach saliva to form nitrosamines, which several studies have shown to be cancer-causing in laboratory animals.[45] A direct link from nitrosamines to cancer in humans has not yet been established.

Metals (heavy)—The human body needs small quantities of some metals, such as iron and copper, to function. But others, especially the so-called "heavy" metals, can seriously interfere with body functions, by displacing the beneficial metals, preventing them from doing their jobs. Best known of these heavy metals are mercury, cadmium and lead. Ingestion of mercury, to give one example, was the cause of the infamous and deadly Minimata disease, which devastated the city of Minimata, Japan, killing scores of people and leaving others blind, deaf, paralyzed or brain-damaged. They had eaten fish and shellfish from Minimata Bay, which was contaminated by methylmercury discharged into the water with waste from manufacturing plants.

Cadmium is a suspected carcinogen that also causes kidney damage. Lead can seriously damage the brain and nervous system, kidneys and bone marrow, as well as impair growth and hormonal development.

Mercury continues to be found in fish, while Health Canada data showed that "elevated levels of cadmium were found in foods normally considered essential for a balanced diet, such as vegetables and cereals, as well as in potato chips and french fries. The highest lead concentrations were found in TV dinners and fish burgers, but ranking high were raisins, muffins, peaches, ground beef and wine."[46]

Nanoparticles (Buckyballs) atomically modified organisms (AMOs)—Nano means one thousand millionth, and nanotechnology means creating products from nano-sized raw materials, namely individual molecules or atoms. In the *Star Trek* television series, ship's crews had "replicators," which could take atoms from any source—garbage, chunks of turf, old car parts, aliens—and by rearranging them, turn them into whatever product they chose. The international food industry is actively trying to do the same.

One website devoted to the subject explains:

> *Manufactured products are made from atoms. The properties of those products depend on how those atoms are arranged. If we rearrange the atoms in coal, we can make diamond. If we rearrange the atoms in sand (and add a few other trace elements) we can make*

> *computer chips. If we rearrange the atoms in dirt, water and air, we can make potatoes.*[47]

Sounds like a wonderful thing, a veritable Magic Bullet. But as with most such wonders, there is a flip side. Rushing recklessly into new technologies, without adequate prior testing or time to assess their long-term effects, can be unwise. Take "buckyballs," or "fullerines," engineered, soccer-ball-shaped molecules of carbon (named for their similarity to Buckminster Fuller's geodesic domes, which look like oversized soccer balls). Researchers hope to use them to create new drugs, cosmetics and undoubtedly foods.

Unfortunately, the effects of these "miracle molecules" are far from benign, as Dr. Eva Oberdorster told the American Chemical Society:

> *[Oberdorster] described what happened when she exposed nine large-mouth bass to water containing buckyballs at concentrations of 500 parts per billion. After only 48 hours, the researchers found 'severe' damage to brain tissue in the form of 'lipid peroxidation,' a condition leading to the destruction of cell membranes, which has been linked in humans to illnesses such as Alzheimer's disease. Researchers also found chemical markers in the liver indicating inflammation, which suggested a full-body response to the buckyball exposure.*[48]

What might be the long-term result of introducing this technology to food manufacturing? As one Health Canada researchers told this author, "It scares me silly."

Organic contaminants (a.k.a. rot and shit)—It was a major scandal (*le scandale de la charogne*, or carrion scandal) in the 1970s when investigators in Quebec found several meat packers were putting spoiled meat in packages and selling it. Organized crime was reportedly involved, and the rotten meat was shipped to several Canadian provinces, as well as the U.S. These days, it's a rare summer when numerous reports are *not* heard of spoiled meat being sold somewhere in Canada or the U.S., most often meat contaminated with *e. coli* or other well-known bacteria.

The public has become accustomed to it, and few give much thought to what could happen if the spoiled meats in question contain antibiotic resistant strains of bacteria. The potential for large-scale harm is ever-present, especially since much of the contamination is due to routine methods of modern, assembly-line meat packing, the end result of which is to make us—literally—eat shit.

The best-selling author of *Fast Food Nation*, Eric Schlosser, wrote: "A nationwide survey by the USDA found that 7.5 per cent of the ground beef samples taken at processing plants were contaminated with *Salmonella*, 11.7 per cent were contaminated with *Listeria monocytogenes*, 30 per cent were contaminated with *Staphylococcus aureus*, and 53 per cent were contaminated with *Clostridium perfringens* ... Behind them lies a simple explanation for why eating a hamburger can now make you seriously ill: There is shit in the meat."[49]

In Canada in 2015, a spike in outbreaks of *listeria* contamination in processed meats was flagged by a food safety expert, who had advised the government during the earlier 2008 Maple Leaf Foods *listeria* incident that killed 22 people. Rick Holley, of the University of Manitoba, reported that there were five times as many food recalls due to *listeria* in 2015 than the year before. Most of the recalls involved cooked meat and fish products, which indicated the pathogen was introduced in packaging. "What this tells me is that maybe the levels of sanitation application on the part of the food industry are slipping," Holley said.[50]

Pesticides and herbicides—The list of pesticides and herbicides whose residues keep turning up on the meat, vegetables and fruit sold in supermarkets is a long one. The suffix "- cide,"as in homicide, indicates killing, and these chemicals are definitely lethal, at least to insects and weeds. Most consumers are well aware of that, which is why people wash food before cooking or preparing it to eat. But washing can't get rid of all of these unwanted extras. How much is still there, and how much is dangerous to humans? Numerous studies indicate there is enough there to cause a serious threat to health.

According to a recent CNN news report, the top herbicide in the world is glyphosate, the active component in Monsanto's Roundup.[51] Roundup was first marketed in 1974, after DDT was banned, but sales boomed in the 1990s, when the company introduced a new strategy, described by Alexis Baden-Mayer of the Organic Consumers Association:

> *The strategy? Genetically engineer seeds to grow food crops that could tolerate high doses of Roundup. With the introduction of these new GE seeds, farmers could now easily control weeds on their corn, soy, cotton, canola, sugar beets and alfalfa crops—crops that thrived while the weeds around them were wiped out by Roundup.*
>
> *Eager to sell more of its flagship herbicide, Monsanto also encouraged farmers to use Roundup as a dessicant, to dry out all of their crops so they could harvest them faster. So Roundup is now routinely sprayed directly on a host of non-GMO crops, including wheat, barley, oats, canola, flax, peas, lentils, soybeans, dry beans and sugar cane. Between 1996 and 2011, the widespread use of Roundup Ready GMO crops increased herbicide use in the U.S. by 527 million pounds—even though Monsanto claimed its GMO crops would reduce pesticide and herbicide use.*[52]

Baden-Mayer's article included an alphabetical list of the health problems researchers have linked in one way or another with glyphosate, giving sources and research study references for each disorder listed. They included:

> ADHC, Alzheimer's disease, anencephaly (birth defect), autism, birth defects, brain cancer, breast cancer, cancer, celiac disease and gluten intolerance, chronic kidney disease, colitis, depression, diabetes, heart disease, hypothyroidism, inflammatory bowel disease, liver disease, Lou Gehrig's disease (ALS), multiple sclerosis (MLS), non-Hodgkin lymphoma, Parkinson's disease, pregnancy problems (infertility, miscarriages, stillbirths), obesity, reproductive problems, respiratory illnesses.

There was nothing under Z.

Roundup was also branded a "probably carcinogen" by the International Agency for Research on Cancer, the cancer research arm of the World Health Organization (WHO), in a report published in the journal *The Lancet Oncology*.[53]

Recent testing has found glyphosate residues in a wide variety of foods, including human breast milk. Abraxis LLC, a Pennsylvania diagnostics company, "found glyphosate in 41 of 69 honey samples and in 10 of 28 soy sauces."[54] Microbe Inotech Laboratories, of St. Louis, Mo., "detected glyphosate in three of 18 breast milk samples and in six of 40 infant formula samples."[55]

Second after Roundup in CNN's top 10 list was Atrazine. "Its use is controversial," said CNN, "due to widespread contamination in drinking water and its associations with birth defects and menstrual problems when consumed by humans at concentrations below government standards." Not above government standards, but "below."

The U.S. Environmental Working Group (EWG) tested two groups of preschool children in Seattle to see whether eating organic food reduced their exposure to pesticides, such as those belonging to the organophosphorus group, that harm the brain and nervous system of growing organisms. The tests found that children who ate conventionally grown food had concentrations of pesticide residues "six to nine times higher" than those who ate organic foods. Children exposed to high levels of organophosphorus pesticides are at high risk for bone and brain cancer, and for childhood leukemia.[56]

And so on. Report after report, for a host of compounds, tells the same story.

Perhaps most shocking, however, is the fact that, thanks to genetic engineering, some modern foods are not simply dosed with pesticides, but **are themselves pesticides**. For example, a genetically modified, eight-trait corn, called "SmartStax," marketed by Monsanto and Dow AgroSciences "produces six different insecticidal toxins" of its own, as well as being tolerant to sprayed herbicides. It "was allowed onto the market in Canada without a safety evaluation from Health Canada."[57]

"Ninety nine per cent of GMO crops either tolerate *or produce* insecticide," according to another report.[58]

The list of contaminants in factory-farm foods could go on, and on, to include literally thousands of substances, some poisonous, some possibly poisonous, some harmless and some whose effects are simply unknown. It is useful to compare this with our grandparents' day, when most foods contained only two additives: salt and pepper.

d) ***Less flavourful***—Because fruits or vegetables harvested ripe from the vine or tree could—after being transported hundreds or even thousands of miles across the world and displayed on market shelves—be spoiled or rotten, industrial growers can't afford to wait to harvest them when they are naturally ready. Instead, they harvest fruit such as tomatoes at the mature green, or "breaker" stage, when the red is just starting to be visible in the product.

The fruit is then ripened artificially, either in trucks in transit or in special "ripening rooms" at its destination, by treating it with ethylene gas. Sometimes, plants are treated while still in the field, with the multi-use pesticide/plant growth regulator ethephon.[59]

Ethylene gas, also known as the "death" or ripening hormone, is given off naturally by ripening plants, and its presence acts as a catalyst to promote still further ripening. In the field, the process occurs comparatively slowly, giving the fruit plenty of time to form the natural sugars that give it flavour. In trucks or ripening rooms, there isn't enough time to allow this. As a result, "tomatoes picked when fully or partly green or at the breaker stage and artificially ripened may be less sweet, more sour, have more off flavour, and less tomato-like flavour than those left on the vine to ripen."[60]

Not only that, but they may be less nutritious: "vine-ripened fruits are higher in ascorbic acid [vitamin C] than those that are artificially ripened."[61]

As for ethephon, human subjects dosed with it in controlled tests reported "sudden onset of diarrhoea or an urgency of bowel movements, stomach cramps or gas and increased urgency or frequency of urination, and either an increase or decrease in appetite."[62] Hopefully residues won't be present on your next dish of tomatoes.

~

> **Looking at the ever-lengthening list of contaminants, the ever-shrinking list of nutrients (on a graph this would, quite appropriately, form an X) and the gradual elimination of flavour, the logical end-result of the factory farm regime—its output, in system terms—would appear to be the production by robots (human labour having been eliminated altogether) of an inert substance, solid enough to provide the sensation of eating, but devoid of either nutrients or flavour and so crammed with poisons as to be almost instantly lethal.**

If this was all that was happening, it would be bad enough. But it isn't all.

4. ***Environmental effects***

Factory farming affects the environment in a host of ways, nearly all of them negative, as a large body of research has shown. To attempt to describe the destruction in complete detail would require a separate book, so only a few examples will be highlighted.

a) *Water pollution and aquifer depletion*—In addition to the numerous herbicides and pesticides described above, all of which eventually find their way into the groundwater adjacent to factory farms, and subsequently into the rivers and streams nearby, the fertilizer regime of industrial operations creates its own pollution problem. The array of possible health effects of pesticides is nearly matched by this second source of pollution.

For example, concentrated, inorganic mixes of nitrogen, phosphorus and potassium (NPK) are routinely dumped on wide swaths of agricultural land to make up for the rapid loss of soil nutrients caused by monocropping. Nitrates and phosphates not taken up by crops enter the soil water, and from there our water supply. When the nitrogen level in drinking water (as nitrate plus nitrite) exceeds 10 mg/L, it can interfere with oxygen transport within the human body. According to Agriculture Canada:

Children under one year old are particularly susceptible, as are cattle and young animals. There have been cases where excessive nitrate in the water of farm animals has reduced conception rates and decreased the number of live births.[63]

Heavy use of fertilizers also leads to eutrophication (an oversupply of nutrients) in streams, ponds, rivers, lakes and even the oceans (which deserve a separate section: see **Global Warming**, below). This can cause sudden, abundant growths of algae (referred to as algal blooms) and other plants. Living plants take up carbon dioxide and give off oxygen, but in the case of some species of algae, whose life cycles are very short, an overabundance of CO_2 can lead to overproduction of algae, which then die and decompose. Decomposition is like a very slow fire, using up oxygen. The decomposing—or oxidizing—algae absorb oxygen as they rot, disturbing the underwater chemical balance and leaving insufficient oxygen for fish and other creatures. Massive fish die-offs have occurred after such contamination. Dense growths of algae can also emit a variety of toxic compounds, to which fish and livestock drinking from polluted water are vulnerable. Pigs are especially sensitive to algal toxins. Phosphorus stimulates the growth of blue-green algae and agriculture is estimated to contribute approximately 40 to 60 per cent of the phosphorus entering the Great Lakes through tributary rivers.[64]

Animal waste is also a leading pollutant. The wastes from chickens in a large battery hen operation, some of which house more than a million birds, can be equivalent to that from a city of 68,000 people.[65] The infectious bacteria salmonella can survive for up to a year in liquid manure and is easily transmitted to people. According to Agriculture Canada, other infections transmitted to humans via manure include "anthrax, tuleremia, brucellosis, erysipelas, tuberculosis, tetanus and colibacillosis."[66]

As for depletion of freshwater reserves, this has become a major threat for the western U.S. states, especially California, and the world. According to space satellite data, California's Central Valley, considered the fruit and vegetable bowl of the U.S., "lost 20.3 cubic kilometers dirt. And trying to boil all that biochemical complexity down to a few paragraphs would be impossible. Suffice it to say that soil is a teeming ecosystem all its own, in which mineral particles, decaying plant and animal matter, bacteria, viruses, fungi, worms, insects, nematodes, moss, plant roots and water all interact and react to create a fertile place to grow things. A single cup of good soil may contain more microbes than there are humans on planet Earth.

Industrial agriculture destroys all of this, replacing the diverse soil biome with just three main chemicals—nitrogen, phosphorus and potassium (NPK)—in concentrated, inorganic form, and, of course, with toxic pesticide and herbicide compounds. This combination, repeated season after season, year after year, burns most organic life in the soil dead, while simultaneously destroying the soil's physical structure, its friability, or tilth. Crops grown in this medium, which some have likened to outdoor hydroponics, may be abundant in terms of volume and look healthy to the eye, but as this chapter has already shown (above, under "less nutritious") those crops are a pale shadow of what farms once grew.

For example, in order to nourish plants, nitrogen "must first be converted to ammonium or nitrate. Soil microbes, which are critical to the nitrogen cycle, achieve this conversion by feeding on decaying plant matter, digesting the elemental nitrogen in the decayed matter. The newly available nitrogen is taken up by plants, where it becomes available to humans" who eat the plants. "But the overuse of fertilizers leads to nutrients like nitrogen building up beyond the capacity of soil microbes to convert it into usable, absorbable nutrients. Too much nitrogen actually kills plant life."[74]

In addition, pesticides and herbicides, such as glyphosate and glyphosate based formulations (GBFs), "can harm the bacterial ecology of soil and cause micronutrient deficiencies in plants, ... reducing respiration in nitrogen-fixing bacteria"[75] particularly if administered in high dosages. Studies have indicated that part of the problem may be the adjuvants (chemicals which aid the effect of the main ingredient) with which glyphosate is combined in some products, such as Monsanto's Roundup. A 2012 study of Roundup's effect on microorganisms used in dairy products found the formulation had "a microbicide effect at concentrations lower than those recommended in agriculture."[76]

In order to create the massive open fields typical of monoculture, factory farms clear away fence rows and fill in drainage ditches to consolidate parcels of land. This eliminates windbreaks that would otherwise have prevented topsoil erosion by wind. The heavy machinery used to till and seed these large fields also tends to compact the soil under the surface, just below the depth of the plough blade, creating a hard pan that decreases proper drainage and prevents roots from extending deeply enough to reach nutrients.

The combination of burning soil life, compacting subsoil and blowing away topsoil is steadily destroying the world's prime farmland. Topsoil is being swept and washed away 10 to 40 times faster than it is being replenished, according to a Cornell University study published in 2006.[77] Cropland "is shrinking by more than 10 million hectares (almost 37,000 square miles) a year due to soil erosion ... The U.S. is losing soil 10 times faster—and China and India are losing soil 30 to 40 times faster—than the natural replenishment rate." In the past 40 years, more than "30 per cent of the world's arable land has become unproductive."

Over-watering of land via irrigation, a common problem on factory farms, also leads to soil salinization, by raising the water table and causing soluble salts in the earth to rise with it. When the water evaporates, or is taken up by crops, the salts are left in the surface layers, eventually threatening crop survival. California's Salton Sea, which collects drainage water from farms in the Imperial Valley and serves as an "evaporation pond" for the region, takes in more than four million tons of dissolved salts per year and is now 25 per cent saltier than the Pacific ocean. Salinization of groundwater, streams and ponds can lead to fish die-offs.

Once the potential effects of climate change are factored into the overall soil picture, the situation could become catastrophic.

d) ***Wildlife habitat and species loss***—Industrial farming impacts nearly all wildlife, including mammals, birds and insects, first by destroying their habitat and second by poisoning them directly with its wide array of lethal chemicals.

reductions. The Houston toad is an extreme case in that it is now an endangered species due to destruction of its habitat by glyphosate."[83] "Herbicides can indirectly cause birds' starvation or force them to leave treated areas because the herbicides destroy the habitat used by the birds' prey. In Maine forests, a study of bird and insect populations following application of the herbicide glyphosate to clearcuts showed that both birds and insects were less abundant in treated areas. The abundance of insects remained low for three years after the herbicide treatment."[84]

The damage done to birds is bad enough, but the destruction of bees is still more shocking, and self-defeating. This includes not only domestic honeybees, but such wild species as bumble bees (which include several species of the *Bombus* genus), mason bees, squash bees, leafcutter bees, mining bees, sweat bees, and many other non-bee insect pollinators. A study published in 2013 showed more than half of the wild bee species in the U.S. were lost in the 20th century. The study:

> *[M]ade use of a remarkable record made of plants and pollinators at Carlinville, Illinois between 1888 and 1891 by entomologist Charles Robertson. Scientists combined that with data from 1971-72 and new data from 2009-10 to discover the changes in pollination seen over the century as widespread forest was reduced to the fragments that remain today. They found that half of the 109 bee species recorded by Robertson had been lost and there had been a serious degradation of the pollination provided by the remaining wild insects.*[85]

The damage being done to honeybees is the best documented, however, and has been going on for decades. Several mechanisms are at work, including one that illustrates the Systems Theory principles that "you can never do just one thing," and "every solution creates new problems."[86] As Joe Rowland, an international beekeeping consultant, told the New York State Assembly in 2000:

> *... bees in the U.S. are increasingly afflicted with a strain of antibiotic resistant American Foulbrood (AFB). Before the advent of*

antibiotics, this bacterial infection was the most serious bee disease in the world. Tetracycline had been used effectively against AFB for 40 years until 1996. In that year, tetracycline resistance was confirmed in both Argentina and the upper Midwestern states of Wisconsin and Minnesota. Since then, it has spread to at least 17 states, including New York. During the 1990s, millions of acres of Roundup Ready crops were planted in the U.S. and Argentina. According to my information, the antibiotic resistant gene used in the creation of Roundup Ready crops was resistant to tetracycline. After 40 years of effective usage against an infective bacterium found in the guts of honeybees, suddenly two geographically isolated countries develop tetracycline resistance simultaneously. A common thread between the U.S. and Argentina is the widespread and recent cultivation of GM crops containing tetracycline resistant genes.[87]

So in making plants resistant to an herbicide, industrial farmers may also have made a leading bacterial killer of the bees, needed to pollinate those plants, resistant to antibiotics. As systems theorist Draper Kauffman says, "You can never do just one thing."[88]

A second group of pesticides, almost as common as Roundup, are the so-called neonicotinoids, which have been implicated in several studies as a cause of major bee die-offs. These chemicals are used as seed dressings and are taken up and spread to every part of the plants they are designed to protect. Lethal to bees in high doses, they appear to be toxic even at lower levels.

Two studies published in 2012, one conducted in England and one in France, "implicated sub-lethal doses of the pesticides with increases in disappeared bees and crashes in the number of queens produced in colonies."[89] As high as an 85 per cent loss was recorded in the number of queens produced. A third study the same year showed "the pesticides can cause colony collapse disorder (CCD), the name given to the ghostly hives from which bees have vanished."

"This data, both ours and others, right now merits a global ban," said Chensheng Lu of the Department of Environmental Health at Harvard University, who led the CCD study.

Neonicotinoids in sub-lethal concentrations are suspected of altering the immune system function in bees, making them more vulnerable to parasites such as *Nosema ceranae*[90] and the more common *Varroa destructor* mites.[91] The USDA has called *Varroa* "the single most detrimental pest of honey bees."[92]

In 2013, the European Food Safety Authority concluded that three common neocotinoids—thiamethoxam, clothianidin and imidacloprid—posed an unacceptable risk to bees. As a result the European Commission voted to ban use of the trio for two years on flowering crops on which bees feed. They continue to be used in North America and Latin America.

Another factor influencing bee resistance to parasites and bacterial disease is declining genetic diversity. Commercial beekeepers in major industrial agriculture regions typically transport bee hives on flatbed trucks, from factory farm to factory farm, keeping the hives at one location long enough to do the pollination work, then moving on to another farm. Thus fewer hives are required to do the work, and the hives that remain may have less inbred resistance to disease. They may also carry diseases from one area to another.

According to estimates published in 2013, "one in three honeybee colonies in the U.S. died or disappeared last winter, an unsustainable decline that threatens the nation's food supply. Multiple factors—pesticides, fungicides, parasites, viruses and malnutrition—are believed to cause the losses, which were officially announced by a consortium of academic researchers, beekeepers and Department of Agriculture scientists."[93]

All of the causes cited, as well as genetic decline, have a single common source: the industrial agriculture system.

Finally, in addition to the birds and bees, there is the destruction of a group of creatures which may not be as economically important as bees (although they are also pollinators of crops), but in terms of sheer beauty and popular affection, may be just as valuable: butterflies. And of all the butterflies of North America, one in particular stands out, the very symbol of summer, beloved of all, the Monarch, *Danaus plexippus*.

The Monarchs' annual migration from Canada to Mexico and back is a true wonder of nature, in which the bright orange and black wings of the tiny, feather-light creatures carry them hundreds of miles, through all weather. Heavy logging that began in their Mexican wintering grounds several decades ago posed an initial threat, by depriving them of crucial habitat. However, the Mexican government, aware of the popularity of the migrants, has since curtailed much of the illegal logging.

Then a new danger emerged, in the form of the agrochemicals whose presence virtually defines industrial agriculture. Glyphosate herbicides, like Monsanto's Roundup, and neonicotinoid insecticides, have decimated the monarchs, and could wipe them out altogether across North America.

At the crucial, month-long larval stage, Monarch caterpillars, with their familiar white, yellow and black stripes, feed almost exclusively on the leaves of the common milkweed plant. Without it, they cannot live to spin cocoons and later emerge as butterflies. Yet milkweed, once found everywhere in the countryside, along farm lanes and in the fence rows between fields, as well as in empty lots in the suburbs, is disappearing.

Glyphosate is particularly lethal to it, and the chemical's widespread use "undoubtedly is a major cause" of the precipitous decline in Monarch numbers, according to Iowa State University weed specialist Bob Hartzler.[94] Milkweed is native to the U.S. corn belt, where Roundup Ready crops have come to dominate, and Hartzler's samplings over the years tell a stark story.

Where 51 per cent of the farm fields he tested in 1999 contained milkweed, only eight per cent still hosted the plant in 2009—a decrease of nearly 85 per cent in corn and soybean fields. Hartzler added that farm fields were once hospitable to Monarchs, actually producing 78 times more of the insects than non-agricultural habitats. But no more.[95] Milkweed has been estimated to have disappeared from at least 100 million acres of row crops.

"We're killing off its host plant in the summer and climate change is hitting it during the winter," said University of Ottawa biology professor Jeremy Kerr. "Monarchs live on a 'noxious weed.' They should be as common as dirt, but they're not."[96]

"Industrial agriculture is a lethal combination of methods that is causing the extinction of thousands of species worldwide," says Allison Wilson, Science Director of the Bioscience Resource Project. "It is affecting birds, amphibians, bats and other pollinators besides butterflies. Many ecosystems are staring down the barrel.

"The saddest irony is that, though industrial agriculture experts call their methods 'scientific,' using toxins to kill pests runs contrary to all biological understanding, including the sciences of ecology, of evolution, and of complex systems. The proof is that the very best results in all of agriculture come from farming methods that reject all industrial inputs. Agribusiness would very much like that not to be known."[97]

The ultimate effect of destroying the insects required to pollinate plants is frightening, as Dr. Brian Moench points out: **"The world's entire food chain hangs in the balance, because 90 per cent of native plants require pollinators to survive."**[98] That includes most of the major farm crops, from apples and peaches to alfalfa.

e) *Invasion by exotic species/transgenic organisms*—The discovery by Purdue's Dr. Huber of an entirely new pathogen, hitherto unknown to science, mentioned above under GMO food contamination, highlighted a possible threat to human health. But an equally serious threat could be posed to the environment by the introduction of such new transgenic organisms, which could upset the balance of natural ecosystems.

History is replete with examples of the mischief that can be created when exotic organisms are introduced into environments where they have no natural ecological niche, and often no controlling negative feedback loop, such as local predators or competitor species. The introduction of the European starling (*Sturnus vulgaris*) to New York City in the late 19th century, and of the infamous cane toad (*Bufo marinus*) to Queensland, Australia in the 20th century spring to mind.

The starling, introduced for aesthetic reasons, reproduced at a prodigious rate, displacing and nearly wiping out the native North American bluebird, and creating a nuisance with its huge, noisy flocks, raining bird manure. The cane toad, which is poisonous, was introduced to help control sugarcane beetle, but instead destroyed local Australian wildlife species and reproduced so rapidly as to become a virtual plague, the toads' roadkilled bodies, in thousands, greasing highways and causing automobile accidents, while their poison harmed children who innocently played with them.

A rogue microbe, whether deliberately created in the laboratory or accidentally produced, could prove damaging or even fatal to wildlife or domestic livestock, as well as humans.

f) *Superweed/superinsect invasions/dominance*—One of the greatest ironies of factory farming, occurring alongside all its other negative traits, is the unintended **provocation of resistance in both plant and animal pests. Here, as with many of its other effects, industrial farming creates its own negative feedback loops.** First, by creating vast, unbroken fields of a single monocrop, it invites the worst weeds and pests to an unnaturally welcome habitat, or to a feast. Then, in an effort to repel the invaders who come to its banquet, it creates the conditions for them to eventually become invincible.

Heavy and continuing application of herbicides, such as glyphosates, after initially wiping out most agricultural weeds, eventually leads to the appearance of resistant weed varieties. These Superweeds, like the antibiotic-resistant Superbugs plaguing our medical systems, can no longer be killed by herbicides. Some, in fact appear to thrive on them. Reports the *New York Times*:

> *Farmers sprayed so much Roundup [Monsanto's brand of glyphosate herbicide] that weeds quickly evolved to survive it. "What we're talking about here is Darwinian evolution in the fast-forward," Mike Owen, a weed scientist at Iowa State University, said.*
>
> *Now, Roundup-resistant weeds like horseweed and giant ragweed are forcing farmers to go back to more expensive techniques that they had long ago abandoned. (Eddie) Anderson, the farmer, is wrestling with a particularly tenacious species of glyphosate-resistant pest called Palmer amaranth, or pigweed, whose resistant form began seriously infesting farms in western Tennessee only last year.*

> *Pigweed can grow three inches a day and reach seven feet or more, choking out crops; it is so sturdy that it can damage harvesting equipment. In an attempt to kill the pest before it becomes that big, Mr. Anderson and his neighbours are ploughing their fields and mixing herbicides into the soil.*[99]

Giant pigweed (*Amaranthus palmeri*) can stop combine harvesters and break hand tools. According to one report, "It can grow seven to eight feet tall, withstand withering heat and prolonged droughts, produce thousands of seeds and has a root system that drains nutrients away from crops. If left unchecked, it could take over a field in a year ... just two palmer amaranth plants in every six metre length of cotton row can reduce yield by at least 23 per cent. A single weed plant can produce 450,000 seeds ... [100] Some farmers have had to "hire a migrant crew to remove the weed by hand."

"As long ago as 2010, GM crops were estimated to have taken over 85-91 per cent of the area planted with the three major crops, soybean, corn and cotton, in the U.S., which occupy nearly 171 million acres."[101]

Mother Jones magazine reports that: "Nearly half (49 per cent) of all U.S. farmers surveyed said they have glyphosate-resistant weeds on their farm in 2012 ... 92 per cent of growers in Georgia said they have glyphosate-resistant weeds."[102]

As for insect pests, they are doing the same, in ever increasing numbers. This has hardly been an unforeseen development. As long ago as 1990, entomologist George P. Georghiou, of the University of California, Riverside, was writing:

> *Insecticide resistance continues to increase, having been recorded in at least 504 species of insects and mites, a 13 per cent increase since 1984 ... The total of 504 species is probably an understatement of the problem since many cases of pest control failure undoubtedly remain uninvestigated or unreported. In view of the evolutionary nature of resistance, additional cases are bound to arise as new insecticides are being introduced and selection pressure remains high ... most of the economically important species have by now developed resistance to at least one insecticide ...*

> *Resistance in various species of aphids, especially those transmitting plant virus diseases, remains a serious problem. There are more cases of resistance reported in the green peach aphid, Myzus persicae, than in any other species of agricultural importance. Such resistance has been reported from 31 countries, involving a total of at least 69 different insecticides, representing organochlorines, carbamates and pyrethroids.*[103]

Remember: he was writing this 26 years ago. And he added: "Most of these cases of insect resistance concern high intensity [namely, factory farm] cropping systems."[104]

A more damningly obvious indictment of such self-defeating methods could hardly be imagined.

g) *Global warming*—Finally, the factory farm system exacerbates climate disruption, as *Ecowatch's* Ronnie Cummins reports:

> *Our modern energy-chemical-and genetically modified organism (GMO) intensive industrial food and farming systems are the major cause of man-made global warming ... by doing a full accounting of the fossil fuel consumption and emissions of the entire industrial food and farming cycle, including inputs, equipment, production, processing, distribution, heating, cooling and waste ... the picture is clear—contemporary agriculture is burning up our planet ...*
>
> *Nearly 65 billion animals worldwide, including cows, chickens and pigs, are crammed into Confined Animal Feeding Operations (CAFOs) ... CAFOs contribute directly to global warming by releasing vast amounts of greenhouse gases into the atmosphere—more than the entire global transportation industry ... Indirectly, factory farms [also] contribute to climate disruption by their impact on deforestation and draining of wetlands, and because of the nitrous oxide emissions from huge amounts of pesticides used to the genetically engineered corn and soy fed to animals raised in CAFOs. Nitrous oxide pollution is even worse than methane—200 times more damaging per ton than* CO_2.[105]

Despite their serious effect on climate, however, the most dangerous effect of the excess pollutants pumped into the air and water by factory farms may well turn out to be their impact on the phytoplankton in the earth's oceans. Phytoplankton, the tiny, single-cell plants that float near the water's surface, form the base of the underwater world's food chain, and are also our planet's single greatest source of oxygen, which they release during the process of photosynthesis:

> *Phytoplankton absorb energy from the sun and nutrients from the water to produce their own food. In the process of photosynthesis, phytoplankton release molecular oxygen (O_2) into the water. It is estimated that between 50 and 85 per cent of the world's oxygen is produced via phytoplankton photosynthesis. . The rest is produced via photosynthesis on land by plants. Furthermore, phytoplankton photosynthesis has controlled the atmospheric CO_2/O_2 balance since the early Precambrian eon.*[106]

If the oceans' phytoplankton balance is disturbed, by overproduction and decomposition caused by too much CO_2, or destroyed outright by other pollutants, thus wiping out the base of the aquatic food chain, human life could become decidedly precarious. In fact, all life that depends on oxygen to breathe, and the food chain to eat, could ultimately be at risk.

> **We, the other animals, the birds, the insects, the fish, could all starve, or choke to death, or both.**

You can never do just one thing ...

Pure destruction

The factory farm cluster of systems, as it operates today, is thus destructive of nearly everything it touches. Carried to its logical conclusion, it will destroy our very capacity to produce food, or perhaps our very ability to breathe. Stemming, as it does, from a combination of greed and an attempt to impose our will on the natural world via a series of Quick Fixes and Magic Bullets, its salient positive feedback loop produces money for a small group of very large corporations, and their executive boards, but little else, while its very existence creates a series of negative feedback loops that put a brake on most of our planet's life forms, including ourselves.

A group of 100 professors from Dutch universities, experts in different fields of agriculture and food, published a report in 2010 calling for a radical reform of farming, noting that the "adverse impacts of industrial factory farms on animal welfare, water and crop resources, climate change and human health" were too serious to be ignored.[107]

Olivier De Schutter, United Nations Special Rapporteur on the Right to Food, summed it up: "They [governments] have stood idle while food companies have re-engineered the look, feel and contents of our food; they have ignored the health impacts of the agricultural subsidies handed out by their own ministries."[108] Calling for a worldwide shift from industrial to "agroecological" [namely, organic] farming, he added that "conventional farming relies on expensive inputs, fuels climate change and is not resilient to climatic shocks. It is simply not the best choice anymore."

Yet increasingly, particularly in North America, it is the only choice on offer. As long as it continues to be, Newton's Third Law will continue to function as well, with every frontal assault by corporate farmers upon the natural world coming, as the popular phrase goes "right back atcha" in new and ever more dangerous forms of resistance.

In 2019-2020, the outbreak of a coronavirus variety that infects humans caused an international epidemic, whose limits no one can yet predict. In this case, it was not factory farming, but the widespread practice in China of selling wildlife, as well as unvaccinated, never immunized stray dogs and cats, for meat in outdoor markets, that was the epidemic's likely source. In North America, pet dogs are routinely immunized against not only rabies, but also against coronavirus. Unvaccinated strays, or wildlife such as ducks, bats or other species might carry it, and may have been the vector that first transmitted the virus to humans in China, where it mutated into its current threatening form, and caused worldwide panic.

But the damage potential of coronavirus, though obviously serious, is nothing compared to the multiple, mass-scale health threats posed by industrial agriculture.

A long time ago, another writer put it bluntly: "Be not deceived; God is not mocked: for whatsoever a man soweth, that shall he also reap."[109]

---

1 My previous books describe the process in detail: *The End of Food: How the food industry is destroying our food supply and what you can do about it*, (Fort Lee, N.J.: Barricade Books, 2006) and *The War in the Country: How the fight to save rural life will shape our future* (Vancouver, B.C.: Greystone Books, 2009).

2 Frank D. Gardner, *Successful Farming*, (L.T. Meyers, 1916), reprinted as *Traditional American Farming Techniques* (Guilford, Connecticut: The Lyons Press, an imprint of The Globe Pequot Press, 2001), 172.

3 Andre Dollinger, "An Introduction to the History and Culture of Pharaonic Egypt: Apologia and bibliography," personal, online website, as posted 10 March 2015 at www.reshafim.org.il/ad/egypt/economy/index.html 1. This is one of the best websites in existence devoted to Egyptology.

4 Ellen Meiksins Wood, "The Agrarian Origins of Capitalism," *Monthly Review*, July-August 1998, Volume 50, Number 3, as posted online 5 March 2015, at http://monthlyreview.org/1998/07/01/the-agrarian-origins-of-capitalism/ 2.

5 Agriculture in the Classroom, "Growing a Nation: the Story of American Agriculture, Historical Timeline–Farmers and the Land," as posted online 10 March 2015, at www.agclassroom.org/gan/timeline/farmers_land.htm 1-5.

6 *Loc. Cit.*

7 Meiksins Wood, *Op. Cit.*, 5.

8 *Loc. Cit.*

9 *Op. Cit.*, 8.

10 Wikipedia, The Free Encyclopedia, *History of Capitalism*, as posted online 5 March 2015, at http://en.wikipedia.org/wiki/History_of_capitalism 2.

11 Wikipedia, The Free Encyclopedia, *Shay's Rebellion*, as posted online 10 March 2015 at http://en.wikipedia.org/wiki/Shays'_Rebellion 2.

12 *Loc. Cit.*

13 David P. Szatmary, *Shays' Rebellion: the Making of an Agrarian Insurrection*, (Amherst, Massachusetts: The University of Massachusetts Press, 1980, xiv.

14 Wikipedia, The Free Encyclopedia, *History of Agriculture*, as posted online 13 January 2015, at http://en.wikipedia.org/wiki/History_of_Agriculture 12-13.

15 *Loc. Cit.*

16 Wikipedia, The Free Encyclopedia, *Luddite*, as posted online 14 March 2015, at http://wikipedia.org/wiki/Luddite 1.

17 *Webster's College Dictionary*, (New York: Random House, 1995).

18 Wikipedia, *Luddite*, *Op. Cit.*

19 Similar movements, spurred by changes in international trade, were seen in the so-called Third World in the late 20th century, and continue even today, as landless peasants converge on Brazil's notorious *favelas*, or Kenya's infamous Kibera Estates, near Nairobi. This author became personally familiar with Kibera Estates while working in Africa in 1989. Most Kibera residents were refugees from economically nonviable *shambas*, or small farms, in western Kenya, from which

they were either displaced by rich landlords, or which they were trying to keep afloat with cash earned through menial jobs in the city.

20 Tim Lang and Colin Hines, "A disaster for the environment, rural economies, food quality and food security," *Ceres 151*, January-February 1995, 20-1.
21 *Op. Cit.*, 19.
22 See endnote 1, re: Pawlick, *The End of Food* and *The War in the Country.*
23 Mississippi State University, College of Agricultural and Life Science, Agro-Eco-system Information Systems, www.ais.misstate.edu/AEE/2613/cases/diffusion-case.html. Case Study #2, 1-5.
24 Kim Eckart, "If there are no new farmers, who will grow our food?", *Yes* magazine, 7 February 2016, as posted online at www.truth-out.org/news/item/34734-if-there-are-no-new-farmer ... 1.
25 Karen Davis, "The battery hen: her life is not for the birds," as cited on all-creatures.org homepage, www.all-creatures.org/articles/egg-battery.html. 21 September 2004, 1-4.
26 Pawlick, *End of Food*, 8-10.
27 *Loc. Cit.*
28 *Loc. Cit.*
29 Andrew Kimbrell, ed., *Fatal Harvest: the tragedy of industrial agriculture*, (Washington: Island Press, 2002), 79.
30 *Ibid.*, 58.
31 Charles Siebert, "Food Ark," *National Geographic Magazine*, July 2011, Graphic "Our dwindling food variety," as posted online 29 March 2015, at http://ngm.nationalgeographic.com/2011/07/food-ark/food-variety-graphic
32 Dr. Mercola, "Scientists vigorously objected to this food - Are you eating it?", 4, as posted 2 April 2011 at *Mercola.com,*
33 Margaret Munro, "Canadian GM canola escapes into wild: study," *The Saskatoon Star Phoenix*, 6 October 2011, 1, as posted 8 October 2011 at www.thestarphoenix.com/story_print.html?id=5509691&sponsor=
34 United States Department of Agriculture, *National Nutrient Database for Standard Reference, Release 15*, (Washing D.C.: U.S. Government Printing Office, 2002.).
35 *Op. Cit.*, 27.
36 *Loc. Cit.*
37 Kerstin Lindgren, "Cow-free milk: a false solution to industry agriculture," *Truthout*, 28 March 2015, as posted online at www.truth-out.org/opinion/item/29860-cow-free-milk-a-false-sol ... 28 March 2015.
38 Andre Picard, "Today's fruits and vegetables lack yesterday's nutrition," *The Globe & Mail*, 6 July 2002, A1.
39 Donald Venes, ed., *Taber's Cyclopedic Medical Dictionary*, (Philadelphia: F.A. Davis Company, 2001), 601.
40 Turning Point Project, "Unlabeled, Untested and You're Eating It," advertisement No. 2 in a series on genetic engineering, 14 October 1999.
41 *Ibid.*
42 Lucy Sharratt, Canadian Biotechnology Action Network (CBAN), "Unprecedented safety study finds harm from GM corn," news release published online 20 September 2012 at http://sn111w.snt111.mail.live.com/mail/PrintMessages.aspx?cpids=f3 ...
43 Mike Ludwig, "USDA approved Monsanto alfalfa despite warnings of new pathogen discovered in genetically engineered crops," *Truthout*, 25 February 2011, as posted online at http://www.truth-out.org/usda-approved-monsanto-alfalfa-despite-war ... 1
44 *Op. Cit.*, 2.

45 Center for Science in the Public Interest (CSPI), "Chemical Cuisine: CSPI's Guide to Food Additives," as posted at www.cspinet.org/reports/chemcuisine.htm 19 January 2001, 20. Also, Ruth Winter, *A Consumer's Dictionary of Food Additives*, (New York: Three Rivers Press, 2009), 382-85.
46 Martin Mittelstaedt, "Canada's food rich in heavy metals, group says," *The Globe & Mail*, 5 May 2003, A1.
47 Ralph Merkle, "Nanotechnology," *Zyvex* website, www.zyvex.com/nano as posted online 9 July 2004, 1.
48 ETC Group discussion list, "Tenth Toxic Warning: more evidence to support nano-moratorium," as posted 1 April 2004.
49 Eric Schlosser, *Fast Food Nation*, (New York: Perennial, 2002), 197.
50 Robin Levinson King, "Rise in recalls due to listeria cause for concern, scientist says," *The Toronto Star*, 2 April 2015, as posted online at www.thestar.com/news/canada/2015/04/02/rise-in-recalls-due-to ...
51 CNN, Plant Protection, "Top 10 classic herbicides used in USA and all over the world," as posted online 26 March 2015 at http://allplantprotection.blogspot.ca/2012/04/top-10-classic-herbicide ...
52 Alexis Baden-Mayer, "Monsanto's Roundup: enough to make you sick," *Truthout*, 1 February 2015, as posted at www.truth-out.org/news/item/28856-monsanto-s-roundup-enough ...
53 Tracy loew, "Roundup a 'probable carcinogen,' WHO report says," *USA Today*, 20 March 2015, as posted online at www.usatoday.com/story/news/nation/2015/03/20/roundup-prob ...
54 Carey Gillam, "Fears over Roundup herbicide residues prompt private testing," *Reuters*, 10 April 2015, as posted online at www.reuters.com/article/2015/04/10/us-food-agriculture-glypho ... 11 April 2015.
55 *Ibid.*
56 Mark Witten, "On the Healthy Promise of Organic Foods," *CSL*, May/June 2003, 37.
57 Lucy Sharratt, "Risks of Monsanto's eight-trait GM SmartStax corn unexamined: new report from Europe uncovers remaining safety questions," *cban e-News*, Canadian Biotechnology Action Network, 28 June 2011.
58 Randy Ananda, "Three approved GMOs linked to organ damage," *Truthout*, 8 January 2010, as posted online at http://truth-out.org/archive/component/k2/item/87545:three-approved- ... 27 March 2015.
59 Ahmad Mozafar, *Plant Vitamins: agronomic, physiological and nutritional aspects*, (Boca Raton, Fla.: CRC Press, 1994), 253.
60 Mozafar, *Op. Cit.*, 255.
61 *Ibid.*, 253.
62 Extoxnet Extension Technology Network, "Ethephon," as posted at pmep.cce.cornell.edu/profiles/extoxnet/dienochlor-glyphosphate 28 August 2004, 1-4.
63 K. D. Switzer-Howse and D.R. Coote, "Agricultural practices and environmental conservation," Ottawa: Agriculture Canada, 1984, 12.
64 Switzer-Howse and Coote, "Agricultural practices," 12.
65 Paul E. Taiganides, "The animal waste disposal problem," *Agriculture and the Quality of our environment*, ed. Nyle C. Brady (Washington, D.C.: American Association for the Advancement of Science, 1967), 389-90.
66 Switzer-Howse and Coote, "Agricultural Practices," 13.
67 Sandra Postel, "California farmers go deep into water debt during drought," *National Geographic* Freshwater Initiative blog, 28 February 2011, as posted online at http://voices.nationalgeographic.com/2011/02/28/california_farmers_g ...
68 Richard Oppenlander, "Freshwater depletion: realities of choice," *Comfortably*

Unaware Blog, as posted 29 March 2015 at http://comfortablyunaware.com/blog/freshwater-depletion-realities-of ... 2-3.

69 Oppenlander, *Op. Cit.*, 5.

70 Katy Daigle, "Water crisis coming in 15 years unless the world acts now, UN report warns," *The Associated Press*, 20 March 2015.

71 Bill Paton, "The smell of domestic pig production on the Canadian prairies," in *Beyond Factory Farming: corporate hog barns and and the threat to public health, the environment and rural communities*, ed. Alexander M. Ervin et al. (Saskatoon: Canadian Centrer for Policy Alternatibves, 2003), 80-1.w

72 A. Dennis McBride, State Health Director, "Medical evaluation and risk assessment: the association of health effects with exposure to odors from hog farm operations," North Carolina Department of Health and Human Services, 7 December 1998, as posted online at www.epi.state.nc.us/epi/mera/ilodoreffects.html

73 *Ibid.*, 4.

74 Hannah Bewsey and Katherine Paul, "Four ways industrial ag is destroying the soil–and your health," *Truthout*, 14 September 2014, as posted online at http://www.truth-out.org/news/item/26164-four-ways-industrial-ag-is-d ...

75 Wikipedia, The Free Encyclopedia, *Glyphosate*, as posted online at http://en.wikipedia.org/wiki/Glyphosate, 1 April 2015, 5.

76 *Loc. Cit.*

77 Susan S. Lang, "Slow, insidious soil erosion threatens human health and welfare as well as the environment, Cornell study asserts," *Cornell Chronicle*, 23 March 2015, 1-2.

78 Bridget Stutchbury, *Silence of the Songbirds*, (Toronto: Harper Collins Publishers Ltd., 2007).

79 *Ibid.*, 118.

80 *Ibid.*, 119.

81 *Ibid.*, 114.

82 *Ibid.*, 107.

83 Pesticides Action Network, "Glyphosate fact sheet," *Pesticides News*, No. 33, September 1996, 28-29, as posted online at http://www.pan-uk.org/pestnews/Actives/glyphosa.htm 1 April 2015.

84 Caroline Cox, "Pesticides and birds: from DDT to today's poisons," *Journal of Pesticide Reform*, Vol. 11, No. 4, winter 1991, 2-6, as posted online at http://eap.mcgill.ca/MagRack/JPR/JPR_14.htm 1 April 2015.

85 Damian Carrington, "Loss of wild pollinators a serious threat to crop yields, study finds," *The Guardian*, 28 February 2013, as posted online at www.theguardian.com/uk

86 Kauffman, *Op. Cit.*, 38-9.

87 Leslie Berliant, "Pollution, pesticides and GM crops killing bees?" *Celsias.com*, 2 April 2011, as posted at http://webcache.googleusercontent.com/search?q=cache:P6nxkdQ02EJ:www.celsias.com/a ... 4 June 2011.

88 Kauffman, *Op. Cit.*, 38.

89 Damian Carrington, "Grave threat of pesticides to bees' billion-pound bonanza is now clear," *The Guardian*, 11 April 2012, as posted online at http://www.guardian.co.uk/environment/damian-carrington-blog/2012/ ...

90 Pettis, J.S., vanEngelsdorp, D., Johnson, J. And Dively, G., "Pesticide exposure in honey bees results in increased levels of the gut pathogen Nosema," *Naturwissenschaften*, February 2012, vol. 99(2), 153-8.

91 Gregorc, A., Evans, J.D., Scharf, M., Ellis, J.D., "Gene expression in honey bee (Apis mellifera) larvae exposed to pesticides and Varroa mites (Varroa destructor)," *Journal of Insect Physiology*, August 2012, vol. 58(8), 1042-9.

92 Brandon Keim, "One-third of U.S. honeybee colonies died last winter, threatening

food supply," *Wired Science*, as posted online at www.wired.com/wired science/2013/05/winter-honeybee-losses/ , 3.

93 *Ibid.*, 1,

94 Stu Ellis, University of Illinois, "Monarch butterflies: the victim of unintended consequences," *We Love Butterflies*, 29 November 2010, as posted online at http://welovebutterflies.com/2010/11/monarch-butterflies-the-victim-of-unintended-consequences/

95 *Ibid.*

96 Paul Hunter, "eButterfly program turns insect lovers into citizen scientists," *Toronto Star*, 22 May 2013, as posted online at www.thestar.com/news/insight/2013/05/22/ebutterfly_program_t ..., p.3.

97 Jonathan Latham, "USDA research links pesticides to monarch butterfly declines," *Naturalblaze.com*, 5 April 2015, 1-2, as posted online 6 February 2016 at www.naturalblaze.com/2015/04/usda-research-links-pesticides-t ...

98 Brian Moench, "The autism epidemic and disappearing bees: a common denominator?" *Truthout*, 21 April 2012, as posted online at http://truth-out.org/news/item/8586-the-autism-epidemic-and-disappear ... 2.

99 William Neuman and Andrew Pollack, "Farmers cope with Roundup-resistant weeds," *New York Times*, 3 May 2010, as posted online 4 July 2011 at www.nytimes.com/2010/05/04/business/energy-environment/04 ... 1.

100 Institute of Science in Society, "GMO crops super-weed timebomb explodes," *Wordpress*, as posted online 1 February 2010, at http://laudyms.wordpress.com/2010/02/01/gmo-crops-produce-super ...

101 *Loc. Cit.*

102 Tom Philpott, "Nearly half of all U.S. farms now have superweeds," *Mother Jones*, 6 February 2013, as posted online at www.motherjones.com/print/215546 6 February 2013, 1.

103 George P. Georghiou, "Overview of insecticide resistance," in *ACS Symposium Series 421, Managing Resistance to Agrochemicals, from Fundamental Research to Practical Strategies*, (Washington, DC: American Chemical Society, 1990), 18-19.

104 *Op. Cit.*, 19.

105 Ronnie Cummins, "How factory farming contributes to global warming," *Ecowatch*, 21 January 2013, as posted online 6 February 2016 at www.nationofchange.org/how-factory-farming-contributes-globa ... 1-2.

106 Wikipedia, the Free Encyclopedia, *Phytoplankton*, as posted online at https:en..wikipedia.org/wiki/Phytoplankton, 7 January 2017, 2.

107 Compassion in World Farming, "Call for ban on factory farming by Dutch professors," 17 May 2010, as posted online at www.ciwf.org.uk/news/2010/05/call-for-ban-on-factory-farming ... 1.

108 Olivier De Schutter, "Obesity epidemic endangers health of millions," *Toronto Star*, 7 March 2012, as posted online at www.thestar.com/opinion/editorialopinion/2012/03/07/obesity_ ... 1.

109 *The Holy Bible*, Galatians 6:7.

Chapter Three

Debt sentence

"Saint Peter don't you call me, cause I can't go.
I owe my soul to the company store."
—*Sixteen Tons*, Tennessee Ernie Ford

Filial cannibalism, the practice of killing and eating one's own or one's neighbour's young, is relatively rare among mammals. It usually crops up in desperate cases, where a species is facing starvation, or in situations where genetic dominance is the evolutionary goal.

Rodent species that reproduce very quickly, like mice, voles or Norway rats, may occasionally make too much whoopee, and the resulting overproduction and crowding may lead to hungry, over-stressed individuals eating the results. Male African lions, when they take over a new pride of females, may kill off the young of any rival males, to assure their own leadership dominance.

However, for animals to destroy not only the young of leadership rivals, but their own offspring, and not only in times of scarcity but in times of seeming plenty and abundant resources, is almost unheard of. Except, that is, in Canada, the U.S.A., and to some extent in Britain, where humans, employing government and corporate structures, have created—and barely attempted to contain—a phenomenon called the Student Loan System.

Morphing through various forms over the years, this entity, starting from small, almost innocuous beginnings, eventually grew into the sort of creature normally encountered only in science fiction: an invisible, yet monstrous devourer of all things educational, and especially of those seeking an education. It has tainted the nature of our post-secondary institutions, altering their budgets, their curricula, and the philosophies on which they were originally based. Characterized by privatization, it contributed to a takeover by for-profit companies not only of schools, but of some of government's basic functions, including student loan administration.

It has helped skew the structure of society, and ruin the lives of more than one student generation by forcing them into debt peonage. Some of its predations have been slightly moderated, at least in Canada. But if education is supposed to be the "activities by which a human group transmits to its descendants a body of knowledge and skills and a moral code which enable the group to subsist,"[1] the student loan system continues to function as the opposite, working to financially cripple our descendants, and through them the group.

And this appetite for eating our young hasn't stopped with the classroom, or at the borders of the English-speaking world. It has spread to other areas and economies, even to countries with more enlightened approaches to schooling, creating numerous other, non-educational forms of pain.

Nor is the urge to devour limited only to youthful prey. Our neighbours, of all ages, are also vulnerable (see **Other horrors**, below).

If the purpose of an economic system, as some have defined it, is "to allocate resources in a way which maximizes society's happiness" by allocating those resources "to their highest valued use,"[2] then our corporate-dominated system is missing the mark by a long shot. In fact, since the "indispensible requirement for any economy is that it must provide the material basis for life for a sufficient number of its citizens so that society can continue,"[3] ours is an across-the-board failure.

Young people, however, are the most heavily impacted by it, and student loans, despite some recent tweaks, are still among the worst crosses they bear. Thus it seems only fair to give them pride of place, and start this chapter with student loans, focusing chiefly on Canada as an example (The situation in the U.S. is not much different, except for being on a much larger scale. And a good book has already been published on it, namely *The Student Loan Scam: the most oppressive debt in U.S. history—and how we can fight back*, by Alan Michael Collinge[4]).

Massive problem

The problem is massive. According to the Canadian Association of University Teachers, there were more than 1.7 million students enrolled in Canada's universities and community colleges in 2008/2009,[5] and the number has only grown since. But getting a numerical fix on their financial situation can be difficult. For example, the Canadian Federation of Students has reported that:

> *In September 2010 the total amount of student loans owed to the [federal] government reached $15 billion, the legislative ceiling set by the Canada Student Financial Assistance Act. In response, the government altered the definition of 'student loan' to exclude over $1.5 billion in federal student debt ... In addition, the $15 billion figure actually only accounts for a portion of Canada's total education-related debt, as it does not include provincial and personal loans, lines of credit and credit card debt.*[6]

According to the federation, the federal Canada Student Loan Program (CSLP) alone expected "to lend approximately $2.1 billion during the 2010-2011 academic year." And the rate of loan uptake is projected to rise steeply, from 36 per cent in 2008-09 to 51 per cent in 2033-34.[7] Loans granted through provincial government programs, or by private lenders, add to the total.

As long ago as 2006, more than 59 per cent of students graduating with a bachelor's degree had significant debt burdens, averaging $24,047 per student. The figures are considerably higher today, and the interest being charged on those loans is unsustainable.

A 2007 report by the Coalition for Student Loan Fairness showed that Canadian student loan interest was "more than double [that] charged by most countries."[8] In contrast, Germany and New Zealand charge no interest on student loans.[9] While the CSLP borrowed money to fund its loan program at "an average rate of 4.45 per cent," the report noted "when students repay the loans, starting six months after graduation, they pay between 8.5 and 11 per cent interest." The system thus

profited from its loans. By 2017, federally-financed student loans were offered at an even higher fixed rate of "prime plus 5 per cent [at the time totalling 9.5 per cent]," or a somewhat lower floating rate of "prime plus 2.5 per cent [at the time totalling 7 per cent]."[10] Interest was originally charged from the time a loan was granted. But in 2012, the federal government elected to begin charging it only upon graduation—still six months before students must start repayment.

When the so-called "sub-prime mortgage bubble" burst beyond the bounds of the real estate sector in 2007-2008, it nearly plunged the world into a second Great Depression. What might happen when this student loan bubble, and all the other the looming bubbles (see below), are pierced?

~

It wasn't always thus. Traditionally, in both Canada and the U.S., as well as other economically developed countries, higher education was seen as a public good and most universities have been non-profit enterprises. From 1939 to 1964, government funding of post-secondary education in Canada came under the Dominion-Provincial Student Loan Program, which consisted of matching **grants** (not loans) from federal and provincial governments. This was roughly the same era that saw returning U.S. war veterans going to university on the so-called G.I. Bill (1944),and Canadian veterans under the Veterans Rehabilitation Act (also 1944). Tuition in both countries was much cheaper in those days—average Canadian national tuition in 1951, for example, was only $230 per year—and after the Second World War there had even been talk of scrapping tuition fees altogether.

Although in the Canadian government system education came under provincial jurisdiction, the federal government accounted for a significant share of financial contributions, particularly after the 1951 report of the Royal Commission on National Development in the Arts, Letters and Sciences, chaired by Vincent Massey, which urged federal "patronage" for students at both graduate and undergraduate levels.

The prevailing belief was that post-secondary education should be dictated by students' ability and desire, not their financial means, with government providing the bulk of support for college and university budgets. As late as 1978, government funds accounted for 83.8 per cent of university operating revenue.[11]

But what would prove to be the first crack in the dike had already appeared. In1964, as part of a general upsurge in investment in education, Lester Pearson's Liberal government created the Canada Student Loans Program, to supplement students' private resources. Loans (*not* grants) were provided to qualified students, *not* by government, but by banks and other financial institutions, which also administered the repayment process.

These loans were in turn backed by the federal government, which reimbursed corporate lenders in full for loans that went into default. The arrangement was, obviously, a bonanza for lenders, who had no-risk access to a burgeoning market. The genie of greed started to emerge from its bottle, and throughout the 1960s, 70s and early 1980s, post-secondary education expanded steadily on this model.

Things began really to change for the worse, however, in 1986, when Brian Mulroney's Conservative government limited cash transfers to the provinces for post-secondary schooling to growth in the national economy, minus two per cent. More transfer cuts were made in 1989, and in 1991 funding for Established Programs Financing was frozen altogether.

Under Jean Chretien's Liberals the situation continued to spiral downward. In 1995, the euphemistically-named Student Financial Assistance Act was passed, ending the full federal guaranty to banks for loan defaults, and substituting instead a so-called "risk share" fee of five per cent payable to banks as compensation for loan losses. The next year, 1996, then-Finance Minister Paul Martin grouped all federal cash transfers for post-secondary education, health and social services into a single Canada Health and Social Transfer—and cut the total amount by $7 billion. The federal cutback was accompanied by decreased funding from many provincial governments as well.

In the year 2000, the federal government eliminated the "risk-share" agreement with banks and took back the financing of all new student loans, along with their administration. It assigned collection of defaulted loans to the Canada Revenue Agency. This moderated the private lenders' bonanza somewhat, but schools still had to cover their costs and pay their teachers, so tuition fees mushroomed.

Between 1985, the year before Mulroney changed the transfer formula, and Martin's 1996 slashing of transfer funds, average annual university tuition had already more than doubled, from $1,019 to $2,384. After Martin's transfer cuts, they more than doubled again, to nearly $5,000 by 2010. As of the 2010-11 school year, the national average was $5,138 per year.

Students and their families had to fork out the difference. Government tuition grants to the students of yore were largely history (for example, the Canada Millennium Scholarship Foundation Bursary, or CMS grant, ended in 2008), replaced by loans, while relief in the form of other, private scholarships was barely a drop in the bucket. With deep cuts in government support to universities simultaneously forcing tuition fees upward, students who weren't from wealthy families had no choice but to cover the gap by taking on debt.

And lenders, within and outside the CSLP, were happy to help them go into debt, reaping the interest lotto. Although government had taken back financing of federal student loans *per se*, chartered banks in Canada developed special programs for students in professional majors (such as medicine or law), whose post-graduate earnings might be higher than average, offering lines of credit over and above their student loan amounts.[12] Inexperienced in money management, students were encouraged by some lenders to borrow not only to pay for tuition and books, but for daily living expenses and various consumer items, such as appliances or used cars (see numbers 18, 27 and 29, below). And many were also taking on provincial government loans.

In system terms, the positive feedback loop created by the Pearson government in 1964, whose output was debt, continued in enthusiastic swing.

The same thing, and worse, happened in the U.S.A., where "a stunning 42 million people now owe $1.3 trillion in student debt." Researcher James Steele explained that "if you go back to [President] Lyndon Johnson, the whole idea of the student loan program was to provide a way for people—poor, minorities, so forth—to give them a way to go to college, and it was to provide the loans. And it was administered largely by the federal government, though banks were involved. But by the late 1990s, through the privatization of Sallie Mae [the U.S. government's student loan vehicle], the signal that sent, they turned over this extremely important function, largely, to private industry—banks, other financial institutions, private equity companies—not just the issuing of loans, but also the servicing of the loans and, maybe more importantly, those that collected the delinquent loans."[13]

Road to ruin

"How do I love thee?" sang the poet. "Let me count the ways."[14]

"How do I ruin thee?" the student loan system seems to mock. And, particularly during its pre-2000 heyday, in roughly ascending order of destructiveness, here are some of the ways Canada's method of financing post-secondary education ruined, and in many ways is still ruining young people and the country.[15] Our system, much like its U.S. counterpart:

1) prices post-secondary education, at both community college and university levels, out of the reach of an increasing number of high school graduates;
2) discourages those who earn community college diplomas from going on to university programs;
3) discourages university graduates with bachelor's degrees from going on to graduate school for a masters or doctorate. In some cases, honors graduates who might have contributed uniquely valuable research have dropped out of further education and gone to work at minimum wage jobs, rather than continue to accumulate more debt. Sociologists call this "debt aversion," and agree it makes sense when a typical bachelor's degree graduate already owes $18,000 to $25,000 on leaving school;
4) prices medical school out of the reach of many students from low-income or middle class families, skewing the profile of med school entrants in favour of the wealthy;

5) saddles medical school graduates who are not from wealthy families with average debts of $250,000 or more;
6) encourages medical graduates to practice in affluent urban areas, rather than poorer rural or inner-city areas, so as to earn more money and pay off their loan debts more quickly;
7) encourages medical graduates to avoid family medicine and opt instead for more lucrative medical specialties, so as to pay off loan debts more quickly;
8) encourages some medical graduates to emigrate to the U.S., where the for-profit health systems in some states may yield higher salaries, with which to pay off student loan debts;
9) prices law school out of the reach of many students from low-income or middle class families, skewing the profile of law school entrants in favour of the wealthy;
10) encourages law graduates to practice in affluent urban areas, rather than in rural or inner-city areas, so as to earn more money to pay off student loans;
11) discourages law graduates from doing *pro bono* (free, "for the public good") work, such as taking on some class action suits or doing Legal Aid work;
12) encourages law graduates to specialize in more lucrative corporate or business law, rather than other specialties like environmental law, so as to pay student loans more quickly;
13) forces students with physical or mental disabilities to pay significantly more for their education than non-disabled students. A deaf student, for example, may pay up to $60,000 more for a bachelor's degree than his or her non-deaf counterpart;
14) prevents disabled students from attending post-secondary schools set up specifically to accommodate their disabilities. In some cases, enrolment in such specialty schools has dropped by 70 per cent or more;
15) through "capping" limits on maximum borrowing periods or amounts, (for example, a 400-week cap for undergraduate degrees) interrupts students and can cut off their loans before they have completed a degree. Students working towards masters or

doctoral degrees are especially vulnerable. Their studies can be halted and loan payments and interest become due before they can finish their work;

16) leads to the gouging of foreign students who wish to study in Canada, treating them as "cash cows." This violates the 1976 International Covenant on Economic, Social and Cultural Rights, to which treaty Canada is a signatory;
17) hobbles college and university graduates with heavy debt for years after graduation, preventing them from making other economy-stimulating investments, such as buying homes or autos, as well as from marrying and starting families;
18) tends to encourage students—particularly medical and law school students—to take on more loans than they really need, thus increasing their post-graduation period of indebtedness. Banks often hard-sell such "preferred" students to take on "generous" lines of credit over and above their government loans;
19) is a needlessly complex system, rife with confusion, incompetence, invasion of privacy and fraud. In many cases, borrowers cannot even find out how much they owe, what the interest is on their outstanding balance, what their monthly payments are or how many years it will take to pay off their loan. When figures are given, they often change without explanation, and no evidence is provided as to the accuracy of the amounts;
20) saddles graduates who miss payments, or make late payments, with penalties, fees, collection charges and accrued compound interest, sometimes totalling more than the original loan itself and creating a *de facto* tax on the working poor;
21) until the year 2000, when the government took back loan administration, often forced students—even including some who had not missed or been late for any loan payments—into default, resulting in their being ruthlessly hounded by debt collection agencies;
22) gave rise to a mass of collection agency abuses, including illegal actions, harassment and abuse of third parties (families, neighbours, co-workers or employers of graduates);

23) has given rise to fears that Canada may adopt the U.S. practice of jailing graduates who cannot make student loan payments, reviving the supposedly long-dead concept of debtors' prisons ("Are there no prisons? And the union workhouses, are they still in operation?" railed Dickens' Ebenezer Scrooge[16]);
24) put some private student lending, even after the federal government took back CSLP financing, into the hands of companies, or their subsidiaries, with histories of lawbreaking, on both sides of the Canada/U.S. border;
25) caused the federal government, since 2000, to assign collection of loan defaults to the Canada Revenue Agency, rather than private collectors. The agency, however, recruits personnel from the same private companies it has replaced, personnel whose biases and methods of work may remain after they've changed jobs;
26) has made students and graduates into a new class of citizen, shorn of a basic right enjoyed by all other Canadians. While others, should they fall upon hard times due to sickness, unemployment or family emergency, may clear their debts by declaring bankruptcy, students cannot. Alone among the population, holders of student debts are forbidden to discharge them in bankruptcy for at least seven years after graduation. Many see this as a violation of the Canadian Charter of Rights (equivalent of the U.S. Constitution), while pointing out the irony that badly-managed collection agencies that fall on hard times may declare bankruptcy, but their targets cannot;
27) has led to potential conflicts of interest, where university employees and administrators have contacts or business relationships with companies that provide loans, or with collection agencies. Some schools distribute preferred lender lists to students, naming particular companies. In one case, a university president and board member was also an executive employed by a private loan company;
28) starves university researchers of funds to carry on their work, while simultaneously weeding out many promising students who may have become brilliant researchers;

29) encourages a mindset among university administrators that favours a "pro-business" model, putting profit above academic freedom and diversity, and discouraging the idea of education as a human right, rather than a commodity for sale. In the U.S., for-profit universities have tried to enroll "as many students as possible, with little regard to ability to complete the degree or pay back the loans. Nearly one in four students at these colleges default on their loans within three years."[17]
30) encourages universities to invest in narrowly business-related or technology-related disciplines, rather than in the arts or in pure research disciplines;
31) makes colleges and universities more dependent on donations from private industry, which often come with strings attached;
32) discourages the original goal of university education, namely to develop graduates with the ability to think independently, creatively and critically;
33) encourages the development of timid, reluctant graduates, more preoccupied with keeping a job, so as to pay off loans, than in contributing to society;
34) creates, via privatization, a non-accountable kind of "shadow government," not subject to regulations that bind the regular civil service, which eventually spreads to many other phases of the bureaucracy, not only education. One privately owned student loan service provider, for example, has also taken over the production of a provincial government's publications, as well as much of the administration of the controversial federal long-gun registry;
35) prevents young people—by curtailing their school careers and forcing them into menial jobs to pay off loans—from reaching their full professional earnings potential, thus assuring that they will contribute less to society in terms of lifetime taxes;
36) assures higher unemployment and under-employment rates, by preventing young people from obtaining optimal training for jobs;
37) damages the physical and mental/emotional health of students and graduates, sometimes severely;

38) drives students and graduates with heavy debts into illegal and dangerous activity, to earn money to pay off loans. These activities may include prostitution, gambling, involvement in the drug trade, and taking part in risky medical trials;
39) drives many students to contemplate suicide, and has driven several, in both the U.S. and Canada, to actually kill themselves.

This is only a partial list.

Of course, no system developed by humans is all bad. On the plus side, our way of financing post-secondary education:

1) has increased the profits of several banks, finance agencies, and especially collection agencies, while providing jobs for their staffs, which may number in the hundreds;
2) has given politicians an excuse to claim that their programs are generously "aiding" students (imagine applying the title "auto aid" to the granting of automobile loans with similar terms of credit);
3) has swollen the election campaign coffers of some political parties and politicians;
4) has, in the words of one student wag, enabled neo-conservative ideologues to "realize their wet dreams" by privatizing education.

Life is a trade-off.

Some examples

Going into detail on each of the above points would take more space than what is available for this chapter. Readers who want more in-depth information may wish to look at my 2012 book, *Debt Sentence: How Canada's student loan system is failing young people and the country,*[18] which includes numerous student stories, and first-person narratives. However, a few examples here should provide an idea of the stakes involved. Take points five through eight, on medical student debts, and 10 through 12, on law students:

"I come from a below-average income, middle class family," said a

23-year-old first year medical student at McMaster University in Hamilton, Ontario. "It's a single-parent family. My father died when I was very young. My mom had a job doing clerical work and has a good pension, but her financial position is not particularly strong. A personal dream of mine would be to one day pay off my mother's mortgage." But the student, who already owed $30-36,000 in loan debt when he received his bachelor's degree, expects his debt to mushroom well past $180,000 when he finishes medical school in three years. "That's about the price of a house where I come from," he said.

To make money faster, and more of it, he must consider which medical specialty he wants to focus on, and where he wants to practice. "In terms of specialty, I'm looking at going into something that's shorter, like family medicine, because you can pay the smaller debt off quicker. In other specialties, you're a resident for longer. The other thing, I think I'd be more likely to go to an urban center, even though I'm from a rural area." Work, and pay, is more plentiful in the city, he explains. There's that necessity to service your debt, before you can work in different areas.

A student at McGill University echoed this view. He expects his debt to easily top a "scary" $150,000. "Should something happen to me, if for some reason I can't finish my training or otherwise be employed as a physician, I'll have a huge debt and nothing to show for it." He once thought of going into a specialty called physiatry, physical training and rehabilitation, but dropped the idea because "they make less than family doctors. Physiatrists and psychiatrists both make less than family doctors. They spend a lot of time with their patients. A family doctor will spend maybe 10, 12 minutes with a patient, whereas a psychiatrist might spend half and hour or an hour. It depends on how quickly you can get them through the door.

"I'm a little ashamed to admit it, but why would I spend three or four years extra as a resident doing something to get paid less, when I'm going to have a mountain of debt?"

A census published in 2011 showed nearly a quarter, 24 per cent, of medical students said they planned to choose a specialty with a high earning potential, and 17 per cent said they'd choose a shorter residency program to start paying off their debts sooner.[19] The needs of society,

Money for nothing

Just after the Second World War, when U.S. veterans were taking advantage of the G.I. Bill to go to school, flocks of con-men and grifters began opening "fly-by-night colleges and vocational schools" to "prey upon veterans."[25] The government of the day quickly stepped in to stop the frauds and the vultures disappeared.

But political memories are short. Beginning in 1972, amendments to the U.S. Higher Education Act made it possible once again for profit-making schools to receive government money, including government grants and student loans. Despite a 1980s report contracted by the Secretary of Education, which found "widespread abuses across the industry,"[26] and the admission by a former lobbyist that "35 or 40 per cent of for-profit colleges engaged in fraud,"[27] for-profit schools proliferated. An aggressive lobbying campaign helped the process, as did changes in 1998 to the Higher Education Act that resulted in further deregulation.

By 2009, the industry was thriving, and large institutional investors, such as banks, hedge funds and institutional retirement funds, had become major sources of capital for it. For example, Wells Fargo and Company, one of the world's largest banks, became a major funder of the for-profit Corinthian Colleges, while Wall Street giant Goldman Sachs invested in the Education Management Corporation.[28]

Both private and government money poured in. By the 2009-2010 academic year, for-profit schools accounted for more than 20 per cent of all federal aid, including $32 billion in Higher Education Act Title IV funds.[29] Some private colleges, such as Corinthian Colleges, even issued their own, high-interest student loans. The for-profits were awash in money.

But, as in the past, the industry's sins began to catch up with it. In 2010, a report by the Government Accountability Office (GAO) showed that "misleading sales and marketing techniques [were] used by several for-profits," and industry critics noted that "more than half of for-profits' revenues were either being spent on marketing or extracted as profits, with less than half spent on instruction."[30] The following year, a study by the National Bureau of Economic Research showed students attending for-profits were "more likely to be unemployed, earn less, have higher debt levels, and are more likely to default on their student loans than similar students at non-profit educational institutions."[31]

The bad news continued and increased, with exposés of various shady practices coming in the press and in television documentaries. Lawsuits were filed against schools for using "alleged aggressive and deceptive sales and marketing practices," and an official of the U.S. Consumer Financial Protection Bureau, Holly Petraeus, claimed:

> *... for-profit colleges see [military] service members as nothing more than dollar signs in uniform, and to use aggressive marketing to draw them in and take out private loans ... One of the most egregious reports of questionable marketing involved a college recruiter who visited a Marine barracks at Camp Lejeune, North Carolina. As the PBS program Frontline reported, the recruiter signed up Marines with serious brain injuries. The fact that some of them couldn't remember what courses they were taking was immaterial, as long as they signed on the dotted line.*[32]

The goal of many for-profits was clearly not to educate anyone, but to make money, even by exploiting vulnerable, wounded war veterans. By 2016, researchers at the National Bureau of Economic Research found that students who went to for-profits "would have been better off not going to school."[33] In the same year, Corinthian Colleges was forced to pay more than $1.1 billion to the State of California, "for false advertising and predatory business practices," and "for defrauding thousands of students."[34]

All of this brought the inevitable backlash. For-profit enrolment peaked in 2009, and afterward posted major declines. According to *Wikipedia*:

> *In 2016, the* Wall Street Journal *reported that more than 180 for-profit college campuses had closed between 2014 and 2016. For-profit college enrolment dropped about 15 per cent in 2016, a 165,000 decline. Enrolment at the university of Phoenix chain fell 22 per cent, which was a 70 per cent loss since 2010. DeVry University reported a 23 per cent drop in 2016. Hondros College, a nursing school chain, dropped*

> *14 per cent. In June 2016, Education Management announced it was planning to close all 25 Brown Mackie College campuses. In September 2016, ITTTech closed all of its campuses. In September 2016, the U.S. Department of Education stripped the Accrediting Council for Independent Colleges and Schools (ACICS) of its accreditation powers.*[35]

Meanwhile, in Canada, the most prominent for-profits operated under the Everest Colleges brand, owned by Corinthian Colleges, mentioned above. Their diplomas were "described as worthless as many graduated students found no job placement, the reputation tainted."[36] On 19 February 2015, the Ontario government shut down 14 Everest College of Business, Health Care and Technology campuses, owned by Corinthian. The next day, Everest College in Ontario declared bankruptcy.

Unlike thousands of students, who had already been legally forbidden to declare bankruptcy when they couldn't pay their student loans. The irony was bitter.

The whole history of student loans, as one wag put it, "smells more like a fishmarket than a market for knowledge."

Responding to popular rumblings of discontent over the situation, and perhaps with an eye on earlier Occupy Movement demonstrations in Quebec, the Ontario government in 2017 moved to make things a little easier for students from poorer families, at least at undergraduate level. "We're making the average college or university tuition free for students from families with incomes of less than $50,000," announced MPP (Member of Provincial Parliament) Sophie Kiwala. Adding that "there will also be more generous grants and loans."[37]

Most students would still be faced with high debts, however, and of course Ontario is only one of several provinces.

Other horrors

As noted earlier, student loans aren't the only economic affliction faced by young people, nor are North Americans or the British alone in suffering hardship. Youth are targeted by other horrors, even in countries where higher education is free.

For example, in most of Western Europe and some other parts of the world, such as the states of the former Soviet Union, or a few Latin American countries, such as Argentina, formal education, including initial-level university, is free. In the Nordic states, post-graduate studies are also covered. But many non-student forms of the debt trap still lurk everywhere, ready to snare the unwary young. So do unemployment, and increasingly often, under-employment.

Consumer loans and credit, whether legal or borderline, are spilling red ink across a steadily broader swath of the globe, making the so-called "millennial" generation (born between roughly 1980 and 2000, also known as Generation Y)[38] and all those born after them the most indebted, financially crippled generations in the history of the industrialized countries. And the most underpaid.

An investigation by Britain's *Manchester Guardian* of seven major economies in America and Europe showed:

> *The full scale of the financial route facing millennials is revealed in new data that points to a perfect storm of factors besetting an entire generation of young adults around the world. A combination of debt, joblessness, globalization, demographics and rising house prices is depressing the incomes and prospects of millions of young people across the developed world, resulting in unprecedented inequality between generations ...*
>
> *Where 30 years ago young adults used to earn more than national averages, now in many countries they have slumped to earning as much as 20 per cent below their average compatriots ... It is likely to be the first time in industrialized history, save for periods of war or natural disaster, that the incomes of young adults have fallen so far when compared with the rest of society.*[39]

Their plight is part of a larger, worldwide trend towards low-paid, often only temporary, work, even for those with university degrees. The *Toronto Star* cited the typical case of a 2011 graduate who "since graduating with a bachelor of arts degree from Brock University, has worked a string of unpaid internships and temporary contracts." He told the

"Love money with your whole heart, your whole mind and your whole soul, and gouge your neighbour for yourself."

The pre-recession period was characterized by banks and other financial corporations offering large numbers of so-called "sub-prime" loans (that is, mortgage or other consumer loans issued at interest rates higher than the Prime Rate set by the U.S. federal Reserve) to low-income or already heavily-indebted borrowers who often could not really afford them. The additional interest charged could translate into "tens of thousands of dollars worth of additional interest payments over the life of the loan," thus increasing bank profits, but in the long run risked default.

In some cases, what loan predators did was illegal, or at best unethical. For example, "a classic bait-and-switch method was used by Countrywide Financial, advertising low interest rates for home refinancing. Such loans were written into extensively detailed contracts, and swapped for more expensive products on the day of closing."[50] Employees of a leading wholesale lender said "they were pushed to falsify mortgage documents and then sell the mortgages to Wall Street banks eager to make false profits."[51] Such loans were often "bundled" with other securities to disguise their true nature.

Eventually, the risks became too high, and the defaults and foreclosures began. The vice-chair of the U.S. Federal Reserve, Janet Yellen, put things in system terms: "Once this massive credit crunch hit, it didn't take long before we were in a recession. The recession, in turn, deepened the credit crunch as demand and employment fell, and credit losses of financial institutions surged. Indeed, we have been in the grip of precisely this adverse *feedback loop* [italics mine] for more than a year."[52]

A rash of scandals involving major banks and financial institutions continued after the recession. Executives of Wells Fargo, "the world's second largest bank by market capitalization and the third largest bank in the U.S. by assets"[53] turned out to have been:

> *... profiteering for years by literally forcing their employees to rob the bank's customers ... executives have pushed a high-pressure 'sales culture,' at least since 2009, demanding that front-line employees*

meet extreme quotas of selling myriad unnecessary bank products to common depositors who just wanted a simple checking account. Employees were expected to load each customer with at least eight accounts, and employees were monitored constantly on meeting their quotas—fail and they'd be fired.

The thievery was systemic, and it was not subtle. Half a million customers were secretly issued credit cards they didn't request; fake email accounts for online services were set up without customers' knowledge; debit cards were issued and activated without telling customers; depositors' money was moved from one account to another; signatures were forged—and, of course, Wells Fargo collected fees for all of these bogus transactions, boosting its profits.[54]

This was the same Wells Fargo and Company which, as noted above, had been a major funder of the now-defunct for-profit Corinthian Colleges, owner of Everest College with its "worthless" degrees.

In 2016, not long after the bank was fined $185 million by regulators, bank CEO John G. Stumpf resigned.[55] He had earlier told government investigators that he and other bank higher-ups knew nothing of the fake accounts, which when discovered had resulted in the firing of some 5,300 lower-level employees.

A few weeks after his resignation, a group of Prudential Insurance Company employees filed a lawsuit against the insurance firm, one of the largest in the country, alleging that Wells Fargo employees had "signed up customers for a low-cost Prudential life insurance policy without their knowledge or permission."[56] The suit said "some Prudential insurance products owned by Wells Fargo customers listed obviously fake home addresses on their applications like 'Wells Fargo Drive,' or phoney email addresses such as 'noemail@wellsfargo.com' ... the insurance premium payments may have come from dormant Wells Fargo accounts."[57]

Wall Street giant Goldman Sachs, which had been heavily involved in the 2008-2009 subprime mortgage debacle, and was later bailed out by the U.S. government, was charged in 2010 by the U.S. Securities Exchange Commission (SEC) with fraud and misleading investors. It paid $550 million to settle the case, $300 million to the government and $250 million to investors, while admitting no wrongdoing.[58] A former Goldman Sachs trader, Fabrice Tourre, "nicknamed 'Fabulous Fab,' was found liable to six fraud claims, in connection with "sub-prime mortgage securities that he knew were doomed to fail."[59]

Goldman Sachs was also an investor in a for-profit college group, Education Management Corporation.[60] Facing enrolment declines and major financial troubles in 2014-2015, the group also closed many of its campuses.

Canadian financial institutions have hardly been angels, in comparison. For example, in April 2016, Canada's anti-money-laundering agency, FINTRAC, announced that a major Canadian bank would be fined $1.5 million for a mind-boggling 1,200 breaches of money laundering laws, but didn't name the bank.[61] In March 2017, Manulife Bank admitted to being the culprit, and on the same day, the Bermuda Monetary Authority announced a $2 million fine against a local division of Sun Life Financial for five breaches of money laundering laws.[62]

Such banks and financial institutions are not marginal industry players, but major corporate leaders, whose actions are seen by many as symptomatic of the culture of the entire sector, popularly symbolized by the fictional movie villain, financier Gordon Gekko, who said in the film *Wall Street*: "greed ... is good." And whether their higher-ups knew it or not, reading the list of corporate misdeeds couldn't help but call to mind images of organized crime's loan sharks, who lend quick cash to the poor at high interest rates. Some media commentators began referring to the denizens of the real Wall Street as "banksters."

At least, the banks did not employ the leg-breaking "enforcers" used by mobsters to collect "the vigorish"!

Regulations such as the U.S. Dodd-Frank Act, introduced in the wake of the 2007-08 debacle to curb the worst lender excesses, have been slated for repeal by the Trump administration.[63] Not long after taking office, Trump called for a "review" of all Dodd-Frank regulations. He also asked the U.S. Labour Department "to scrap the so-called fiduciary rule, which requires investors who manage retirement accounts to act in the best interest of their investors—not Wall Street."[64] His Economic

Advisor, former Goldman Sachs second-in-command Gary Cohn, defended the request.[65]

If such brakes on predatory lending are removed, history could repeat, and the world's already-rising consumer debt could not help but rise still more precipitously, while both income inequality and risk rise with it.

Evil twins

Income inequality and debt are the evil twins causing much of the trouble for the young, as well as everyone else. And both are at unsustainable levels.

For example, in 2017 Oxfam reported that only eight billionaires held as much wealth as 50 per cent of the world's entire population, and that inequality was leading to "suppressing wages, as businesses are focused on delivering higher returns to wealthy owners and executives."[66] Not long after, the number of billionaires needed to equal 50 per cent of the world population in wealth was revised to only six, all men.[67] Credit Suisse had already published similar reports in 2015 and 2016, warning that the accelerating disparity of incomes could in itself trigger a recession.[68] The Organization for Economic Cooperation and Development (OECD) has warned that "income inequality in OECD countries is at its highest level for the past half century."[69]

Such inequality has historically had ominous overtones. As the authors of one paper warn: "The economic stratification of society into 'elites' played a significant role in the collapse of other advanced civilizations such as the Roman, Han and Gupta empires."[70]

Democratic U.S. presidential candidate, Senator Bernie Sanders, warned about income inequality in his 2016 campaign, noting that "America now has more wealth and income inequality than any major developed country on earth, and the gap between the very rich and everyone else is wider than at any time since the 1920s."[71] He called it "the great moral issue of our time."[72]

As for household debt, which includes home mortgages, auto loans, student loans and credit card debt, it is closely related to economic crises.

As the *Wikipedia* notes: "A significant rise in the level of this debt coincides historically with many severe economic crises and was a cause of the U.S. and subsequent European economic crises of 2007-2012."[73]

Canadian household debt soared to record levels in 2016, as *Mortgage Broker News* reported: "In its latest national balance sheet, Statistics Canada revealed that household debt nationwide has reached unprecedented heights ... Canadian households owed $1.68 in debt for each dollar of their disposable income," for a total of $1.97 trillion.[74]

The story was the same in the U.S., where the *Huffington Post* reported that "average U.S. household debt has now passed the $90,000 mark,"[75] and in Britain, where *The Guardian* revealed that "The average UK household will owe close to 10,000 pounds in debts such as personal loans, credit cards and overdrafts by the end of 2016, which is a new high in cash terms." *Nearly half* of that "came from student borrowing."[76]

Another British paper, *The Telegraph*, had warned earlier that "[o]n a global level, growth is being steadily drowned under a rising tide of debt, threatening renewed financial crisis, a continued squeeze on living standards and eventual mass default ... The only way the world can keep growing, it would appear, is by piling on debt. Not good, not good at all."[77]

"Not good" could be an understatement, a possibility Dimitri Papadimitriou, of the Levy Economics Institute, has also emphasized.[78] "The debt held by American households is rising ominously," he warned. "And unless our economic policies change, that debt balloon, powered by radical income inequality, is going to become the next bust ... What's emerging is a new sort of speculative bubble, this time based on consumer and corporate credit ... for the vast majority, wages and wealth aren't going up, so we're anticipating that the majority of Americans—the 90 per cent—will once again do what was done before: borrow, and then borrow more.

"Insolvency for the 90 per cent—the overwhelming majority of Americans—has become, in the decade's catch phrase, 'the new normal.' Unsustainable? Of course."

Key Indicators

Income inequality and household debt are obvious indicators of possible future economic crises, but there are others. For example, the onset of unexpected, or unprepared-for natural disturbances can destabilize economies. And, of lesser importance in the view of many economists, but still a factor, is government debt (a.k.a. national or public debt).

A look at the key factors that most historians agree brought on, or at least exacerbated, the Great Depression of the 1930s, and comparing them with today's situation, should be instructive. There is, of course, considerable disagreement among economists over just which causes, and in what order, deserve the greatest blame. A variety of mechanisms have also been suggested for how they actually interacted with each other to cause disaster.

However, most would likely agree with Martin Kelly, of the Beacon Historical Society, who lists the "*Top 5 causes of the Great Depression*"[79] as:

1) ***Stock Market Crash of 1929***—Many believe erroneously that the stock market crash that occurred on Black Tuesday, October 29, 1929 is one and the same with the Great Depression. In fact, it was one of the major causes that led to the Great Depression. Two months after the original crash in October, stockholders had lost more than $40 billion. Even though the stock market began to regain some of its losses, by the end of 1930, it just was not enough and America truly entered what is called the Great Depression.
2) ***Bank Failures***—Throughout the 1930s over 9,000 banks failed. Bank deposits were uninsured and thus as banks failed people simply lost their savings. Surviving banks, unsure of the economic situation and concerned for their own survival, stopped being as willing to create new loans. This exacerbated the situation leading to less and less expenditures.
3) ***Reduction in Purchasing Across the Board***—With the stock market crash and the fears of further economic woes, individuals from all classes stopped purchasing items. This then

led to a reduction in the number of items produced and thus a reduction in the workforce. As people lost their jobs, they were unable to keep up with paying for items they had bought through installment plans and their items were repossessed. More and more inventory began to accumulate. The unemployment rate rose above 25% which meant, of course, even less spending to help alleviate the economic situation.

4) ***American Economic Policy with Europe***—As businesses began failing, the government created the Smoot-Hawley Tariff in 1930 to help protect American companies. This charged a high tax for imports thereby leading to less trade between America and foreign countries along with some economic retaliation.
5) ***Drought Conditions***—While not a direct cause of the Great Depression, the drought that occurred in the Mississippi Valley in 1930 was of such proportions that many could not even pay their taxes or other debts and had to sell their farms for no profit to themselves. The area was nicknamed "The Dust Bowl." This was the topic of John Steinbeck's *The Grapes of Wrath*.

Subscribers to the so-called Austrian School of economics also identify **government, or national debt** (namely, the accumulated annual budget deficits of a given country) as a factor, arguing that "the key cause of the Depression was the [government] expansion of the money supply in the 1920s that led to an unsustainable, credit-driven boom ... Hans Sennholz argued that most boom-and-busts that plagued the American economy ... were generated by government creating a boom through easy money and credit, which was soon followed by the inevitable bust. The spectacular crash of 1929 followed five years of reckless credit expansion by the Federal Reserve System under the Coolidge Administration. The passing of the Sixteenth Amendment, the passage of the Federal Reserve Act, rising government deficits, the passage of the Hawley-Smoot Tariff Act, and the Revenue Act of 1932, exacerbated the crisis, prolonging it."[80]

This hypothesis, however, remains a minority one. Most economists, especially those who agree with the theories of Britain's John Maynard

Keynes, take the opposite view. Keynes held that "to keep people fully employed, governments have to run deficits when the economy is slowing, as the private sector would not invest enough to keep production at the normal level and bring the economy out of recession. Keynesian economists called on governments during times of economic crisis to pick up the slack by increasing government spending and/or cutting taxes."[81]

History appears to have borne him out. Then-U.S. President Franklin Roosevelt pursued a policy of government expansion of public works, farm subsidies and employment schemes such as the Works Progress Administration, and the U.S. economy began slowly to revive. The common view is "that Roosevelt's New Deal policies either caused or accelerated the recovery."[82] The Great Depression, most agree, was finally ended by the stimulus provided by government military spending for the Second World War.

As the *Wikipedia* puts it:

> *The rearmament policies leading up to World War II helped stimulate the economies of Europe in 1937-39. By 1937, unemployment in Britain had fallen to 1.5 million. The mobilization of manpower following the outbreak of war in 1939 ended unemployment.*
>
> *When the United States entered the war in 1941, it finally eliminated the last effects from the Great Depression and brought the U.S. unemployment rate down below 10 per cent. In the U.S. massive war spending doubled economic growth rates, either masking the effects of the Depression or essentially ending the Depression. Businessmen ignored the mounting national debt and heavy new taxes, redoubling their efforts for greater output to take advantage of generous government contracts.*[83]

And how do conditions stack up today?

1) ***Stock market crashes***—As history has repeatedly shown, almost anything could provoke a stock market crash—defined as "a sudden, dramatic decline of stock prices across a significant

cross-section of a stock market ... They often follow speculative stock market bubbles."[84] As the author of an *AOL Finance* article on the topic explained: "When investors as a group start to close their wallets and sellers can't find buyers, prices will drop more generally. This type of action is typical of a pullback in an otherwise healthy market, but when combined with a negative piece of news or data, it can intensify selling."[85]

The *Wikipedia* adds: "Stock market crashes are social phenomena where external economic events combine with crowd behaviour and psychology in *a positive feedback loop* [italics mine] where selling by some market participants drives more market participants to sell."[86]

A single, reckless "Tweet" from someone like U.S. President Trump, attacking a company or group of companies, or announcing some alarming or abrupt government action, could easily be the "negative piece of news or data" required to start a slide, which AOL very aptly describes in system terms, as a "positive feedback loop."

2) ***Bank failures***—Any loss of uninsured savings due to widespread bank failures should in theory be mitigated this time, as most developed countries now have structures that insure bank depositors' savings. In the U.S. this is done through the Federal Deposit Insurance Corporation (FDIC), while in Canada the Canada Deposit Insurance Corporation (CDIC) does the same. This presumes, however, that such structures will be solvent enough to cover truly enormous losses.

   The U.S. Securities Exchange Commission, created in 1934, and the U.S. Banking Act (the "Glass-Steagall Act") of 1933, were designed to prevent a repeat of the Depression's widespread bank failures. However, they were later significantly weakened by a succession of U.S. administrations, including the Carter, Reagan, and Clinton presidencies. Clinton's Gramm-Leach-Bliley Act effectively gutted Glass-Steagall.

More recently, the Dodd-Frank Act of 2010, mentioned earlier, which curbed some lender excesses and was intended to prevent a repeat of the 2007-2008 Recession, has been slated for repeal by the Trump administration.

With U.S. banking regulation thus significantly weakened, and American banks being exposed to greater chances of consumer debt default, bank failures may again be on the American agenda. And what happens to the U.S. economy will inevitably impact most other world economies.

3) ***Reduction in purchasing***—If consumers, particularly the younger generation, reach such a state of debt and precarious employment that they can no longer even consider buying anything (recall the indebted and impoverished students who can no longer afford to buy food), a general slowdown of the commercial economy is inevitable. A slowdown would also reduce available jobs, as companies cut back due to falling sales. Thus yet another positive feedback loop could be established, spiralling downward.

4) ***American policy with Europe***—The tariff war launched by the Smoot-Hawley Act of 1930 curbed trade between the U.S. and Europe, thus further slowing the economies of both regions. The new Trump administration in Washington in 2017 announced plans to institute an "America First" policy, that could significantly cut imports. If this plan includes tariffs, it could see history repeat. Such a prospect became far more likely in 2018, when Trump launched a possible international trade war, initially targeting Canada, but later including other countries.

5) ***Drought conditions***—The Dustbowl of the "Dirty Thirties" was a major factor worsening the Great Depression. As Chapter Four will show, Climate Change, brought on by modern corporate energy policies, is on track to create the Mother of All Dustbowls, as well as floods, wildfires and weather extremes, including hurricanes, typhoons, tornadoes and wild winds. In short, enough natural disturbances to make the world dizzy.

As for **government debt**, as mentioned above, most mainstream observers don't see it as likely to be the direct cause of a major economic crisis in developed countries such as the U.S., though it could well be a complicating factor. This is because national governments, if their debt gets too high, can always avoid default by raising taxes, cutting spending or simply printing more money to take up the slack.

Observers also agree, however, that government debt around the world has never been as high as it is today, and some countries, particularly those in the so-called underdeveloped world of the Global South, are already in trouble. According to figures released in 2016 by the International Monetary Fund (IMF), total government debt worldwide had risen to a record $152 trillion, equivalent to 225 per cent of global gross domestic product (GDP).[87]

Debt to GDP ratio is an accepted way of assessing the significance of a nation's debt, and the IMF figure comes into perspective when it is realized that one of the criteria of admission to the European Union's euro currency is "that an applicant country's debt should not exceed 60 per cent of that country's GDP."[88]

A CIA Factbook table, released in 2013,[89] showed the U.S. topping the world charts even then with a debt to GDP ratio of 73.6 per cent—meaning that it wouldn't qualify for EU membership! The table showed the U.S. as the number one global debtor, accounting for 31.27 per cent of the world's total government debt. And the figure was for federal debt only, not including the debt of individual U.S. state governments. Russian debt, in comparison, was only 12.2 per cent of GDP, and its share of global debt was only 0.55 per cent.

By the end of the Obama presidency in 2016, U.S. government debt had reached $19.6 trillion and its annual debt growth of $1.36 trillion was "the third biggest annual increase ever."[90]

Whatever the current economic orthodoxy might be, such figures cannot help but be disturbing. The debt crisis in Greece, which prompted draconian EU austerity measures, shocked many. And those with an eye to history were prompted to recall the crisis in post-First World War I Germany, when the Weimar Republic found itself with a massive war debt it couldn't pay.

When the conflict began, German Emperor Wilhelm II and the German Parliament agreed to "fund the war entirely by borrowing ... The government believed it would be able to pay off the debt by winning the war, and it would be able to annex resource-rich industrial territory in the west and east. Also, it would be able to impose massive reparations on the defeated allies ... "The strategy backfired when Germany lost the war. The new Weimar Republic was now saddled with a massive war debt that it could not afford. That was made even worse by the fact that it was printing money without the economic resources to back it up."[91]

The German mark sank in international markets, prompting hyperinflation of the currency, and by November 1923, the U.S. dollar was worth a staggering 4,210,500,000,000 German marks![92] German people were using wheelbarrows to carry enough bills to market to buy a loaf of bread.

What might happen if the U.S. government of today, with its huge debt load and crucial connections to every other economy in the world, should find itself forced to default on its obligations, or to take drastic measures, such as sharp tax hikes or printing worthless currency, to avoid it? Could some combination of wars, environmental collapse or social unrest lead to this? And even if government default, as opposed to consumer, household or corporate debt default, isn't likely, what about the other five causes of the Great Depression, listed above? Is another world financial collapse—this time a Greater-Than-Great Depression—possible?

Trying to predict a financial crisis may well be an exercise in futility, as Bank of England monetary policy committee member Gertjan Vlieghe recently told a British Parliamentary committee. "We are probably not going to forecast the next financial crisis, or forecast the next recession," he said. "Our models are just not that good."[93]

But the real surprise would be if a collapse of some kind wasn't in the offing. Any economy that encourages such widespread financial vulnerability as ours does, while creating conditions among the general population—especially the young—which could obviously provoke a depression, is a disaster waiting to happen.

And, much like today's expensive law courts, mentioned above under the effects of the student loan system, the courts of the *Ancien Régime* were inaccessible because "judges required that both parties pay for the costs of the trial (called the *épices*). This effectively put justice out of the reach of all but the wealthy."[99]

Needless to say, ordinary citizens deeply resented their situation, but a spur, or series of spurs was needed to drive them to action. Not surprisingly, two big ones came along: 1) a grain crisis sparked by the so-called Flour War of 1775, and worsened by later harsh winters and poor harvests caused by a strong two-year *El Nino* weather cycle (likely provoked by the large *Laki* volcano eruption in Iceland),[100] and, 2) the bankruptcy of the government, in the wake of its spending on the Seven Years' War and the American Revolution.

One of the *Ancien Régime's* few useful roles was to act as insurer of the nation's food supply, particularly grain, by policing the grain market and its imports and exports. In regions facing bad harvests, the grain police forbade export and supervised the importing of grain from regions with surpluses. In 1774, however, this system was abolished by the Controller-General of Finances, following the then-popular regulatory notion of "*laissez-faire, laissez passer*, meaning leave it alone and let it pass, also known as the invisible hand notion," and the grain market was deregulated.[101]

Speculators began buying up large quantities of grain, during times of good harvests, and storing them to sell at a high price in times of poor harvest. The price of grain spiked, and became hard for many to afford. When the *El Nino* kicked in, in 1788-89, an event known as the Little Ice Age, France not only saw meagre harvests but also sky-high grain prices.

Then came the government's war debt, which carried a high interest rate and which Louis XVI tried to deal with via an increase in taxes, this time including taxes on the nobility. But his attempts to raise taxes failed, thanks to heavy resistance from the nobles and upper bourgeoisie. The government was, effectively, bankrupt.

Something had to give, and did. People took to the streets, including a demonstration in 1789 by crowds of thousands of women protesting bread shortages[102] (eerily reminiscent of 2017's international women's marches), and eventually the famed storming of the Bastille. Louis XVI lost his head—literally—on the guillotine, and the rest was, as they say, history.

The parallels between then and now are obvious, as are, to a slightly lesser extent, those between the run-up to the 1917 Russian Revolution and today's conditions.

Russian Revolution

In 1917, Russia was mired in the First World War, which it was losing badly. But trouble had been brewing long before. The Tsars (Russian for "Caesars," a holdover from the days of the Roman Empire) were absolute autocrats, if anything more arbitrary and heavy-handed towards their people than the pre-1789 French kings had been.

Russia's peasants, who made up the vast majority of the population, had worked the land for centuries as serfs, owned and traded as literal slaves by the landowners, who included the Tsar, the nobles and a few members of the small middle-class. In 1861 they were freed and given tiny plots of land, but only on condition that they pay "redemption" payments to the state, which plunged them deeply in debt. They eked out an existence at subsistence level, and nearly half of peasant families had at least one member who had left the village to find other work.[103] Like many Third World migrants to the industrialized north today, they sent home what little money they could to help their families.

By 1917, even after emancipation, 25 per cent of the nation's arable land was still owned by only 1.5 per cent of the population—chiefly the Tsar and the nobility,[104] whose opulent estates and lifestyles rivalled that of the *Ancien Régime* in France. Attempts at land reform, sometimes led by politically liberal landowners (see Count Leo Tolstoy's famous essay, "What then, must we do?"), were repressed by the Tsar's "land captains."[105]

Meanwhile, in the cities, the working majority, including many former peasants:

> *... had good reason for discontent: overcrowded housing with often deplorable sanitary conditions, long hours of work (on the eve of the war a 10-hour workday, six days a week, was the average and many were working 11-12 hours a day by 1916), constant risk of injury and death from poor safety and sanitary conditions, harsh discipline (not only rules and fines, but foremen's fists), and inadequate wages (made worse after 1914 by steep wartime increases in the cost of living) ... In one 1904 survey, it was found that an average of 16 people shared each apartment in St. Petersburg, with six people per room. There was also no running water, and piles of human waste were a threat to the health of the workers.*[106]

The war brought, first, conscription into the military, which "stripped skilled workers from the cities, who had to be replaced with unskilled peasants." Then, to finance the war, the government began printing millions of rouble notes, and by 1917 inflation had made prices increase to four times what they had been at the war's outbreak in 1914.

> *The main problems were food shortages and rising prices. Inflation dragged incomes down at an alarmingly rapid rate, and shortages made it difficult to buy even what one could afford ... It became increasingly difficult both to afford and actually buy food.*[107]

On top of all this came battlefield disaster. "By the end of October 1916, Russia had lost between 1.6 and 1.8 million soldiers in combat, with an additional two million prisoners of war and one million missing, all making up a total of nearly five million men."[108] Such staggering casualties, and resulting anger at the Tsar, who had taken direct command of the army in 1915, produced a tipping point. Strikes, demonstrations, soldier mutinies, led to widespread chaos. The Imperial Parliament (Duma) took over control and formed a provisional government. The Tsar abdicated (to be executed later, along with most of his family).

Attempts by the provisional government to bring stability failed, leading to the takeover of government In October 1917 by the Bolsheviks, led by Vladimir Lenin. And the age of Communism, to culminate in the dictatorship of Josef Stalin following Lenin's death, had begun.

Rampant inequality, food shortages, destructively expensive wars and a badly managed economy: all the ingredients had been there.

Rise of the Nazis

Adolph Hitler had been a corporal in the First World War, but his front-line combat experiences seemed not to have mellowed him where violence was concerned. His rise, in tandem with that of the National Socialist (Nazi) Party, was to bring even wider, more destructive world warfare, along with the worst anti-humanitarian atrocities yet seen, during the Holocaust.

As mentioned earlier under National Debt, the 1923 hyperinflation during the Weimar Republic had severely weakened the post-First World War German economy, struggling as it was at the time to repay punitive war debts imposed by the victorious Allies. Widespread hardship resulted, and attempts to revalue the currency had only limited impact. Then, starting in 1924, the United States through U.S. banks began lending Germany huge sums. The country was "rebuilt, unemployment was reduced and people began to feel secure" again. "The years 1924 to 1929 became known as the 'Golden Years.'"[109]

But the raging winds of the Great Depression blew out this all-too brief candle of hope. Following the stock market crash of 1929, the U.S. banks called in their loans, and Germany was once again in crisis. By 1933, Germany had to suspend reparation payments. Its economy collapsed, businesses failed and unemployment topped six million. Hardly recovered from their earlier troubles, the panic-stricken German people wanted answers.

Hitler, the rabble-rousing head of what had once been a radical fringe party, had them. Though mocked by Germany's elites as "the little corporal," and even jailed briefly following an attempted coup in 1923, he was a mesmerizing public speaker and knew how to present simplistic solutions that could appeal to the unsophisticated masses. One of these was to find a scapegoat for their troubles, namely the Jews, against whom he railed.

Street violence broke out between Hitler's followers and the German Communists, who had set up a paramilitary group called the *Rotfront* (from *Roter Frontkampferbund*). Battles between the equivalent Nazi group, the SA, and the Rotfront included gunfire, and several people were killed. Frightened at the possibility of a Communist takeover, wealthy businessmen helped finance Hitler's campaigns and Nazi party membership grew.

Cooler heads, who had hoped the violence would discredit the warring factions, were dismayed when the Nazis gained shocking victories in subsequent elections for the Reichstag, after which the Nazis became the second largest party in the country. Eventually, the political crises worsened, as did manoeuvring among politicians, and the German President, Von Hindenberg decided to appoint Hitler Chancellor. The die was cast, and the road to the Second World War opened wide.

There hadn't been an environmental disaster inside Germany, like the American Dustbowl of the 1930s, or a major agricultural failure, but they weren't needed. Such things had already occurred in the United States, worsening the Great Depression (which, as noted, had prompted the U.S. loan withdrawal of 1929), and a stricken America was unlikely to renew lending money to Germany.

Condemned to repeat

The philosopher George Santayana once said: "Those who cannot learn from history are doomed to repeat it."

While the preludes to the Great Depression, and the French, Russian and German revolutionary crises are obviously not precisely identical to our current reality, the similarities are plain, and plainly numerous.

Most of the world's population, especially and increasingly the young, live in conditions of gross social and economic inequality, as burdened by debt as are their respective national governments. They face chronic unemployment, or precarious underemployment. Their banks and financial institutions periodically careen out of control, with bubble-chasing-bubble, and their politicians—unable to learn from even the most recent history (such as the 2008 Recession)—continue to deregulate, periodically removing whatever controls got put in place in the immediate aftermath of the last crisis.

Instead of a stable, ordered society in which individuals can feel reasonably secure in their interactions with each other, they live in an unjust, unstable society that victimizes the vast majority—the so-called "99 per cent"—and tends, as such societies have in the past, towards economic collapse.

National governments, already floundering in debt, continue investing in their militaries and the arms industry, while pursuing seemingly endless brushfire wars in the Middle East and elsewhere, and flirting constantly with the idea of major war between the major powers. U.S. President Trump has stated his intention to increase spending on America's nuclear arsenal, despite its being already the largest in the world. U.S. military spending, totalling $598 billion for fiscal year 2015, accounted for more than half—54 per cent—of the U.S. federal budget,[110] and its resulting deficits. China was the second largest spender, at $145.8 billion, and Russia 4th at $65.6 billion.[111] The U.S. was also the greatest arms exporter, accounting for 31 per cent of the global share.

Growing numbers of people, most dangerously in the United States, are poorly educated and easily prone to believe in "alternative facts," and simplistic, "Magic Bullet" (sometimes literally) solutions to their problems.[112] Political opportunists scan the horizon for scapegoats. As in Germany in the 1930s, their audiences are ripe prey for demagogues (more on this in chapters 6 and 7).

And finally, looming in the background like a gathering storm, dwarfing the Dustbowl, what could become the greatest natural cataclysm since the extinction of the dinosaurs 66 million years ago, waits for us: Climate Change. And that is the subject of the following chapter.

---

1 Beiter, Klaus Dieter, *The Protection of the Right to Education by International Law*, (The Hague: Martinus Nijhoff, 2005), 19.

2 Bob Reed, What is the purpose of an economy?" *Econrules!*, as posted online 7 April 2017, at www.econ.canterbury.ac.nz/personal_pages/bob_reed/econ3003 ...

3 John Medaille, "Chapter IV: the purpose of an economy. What must an economy do?" 18 June 2008, *The Distributist Review*, as posted online 7 April 2017 at http://distributism.blogspot.ca/2008/06/chapter-iv-purpose-of-econom ...

4 Alan Michael Collinge, *The Student Loan Scam: the most oppressive debt in U.S. history–and how we can fight back*, (Boston: Beacon Press, 2009).

5 Canadian Association of University Teachers, CAUT, *Almanac of Post-Secondary Education in Canada 2011-2012* (Ottawa: 2012), 25.
6 Canadian Federation of Students, *Public Education for the Public Good* (Ottawa: 2011), 14.
7 Superintendent of Financial Institutions in Canada, Office of the Chief Actuary, *Actuarial Report on the Canada Student Loans Program, as at 31 July 2009* (Ottawa: Office of the Chief Actuary, 2009), 22.
8 Coalition for Student Loan Fairness, *"Canadian student loan interest is more than double charged by most countries,"* 5 July 2007, as posted online at: www.bankruptcycanada.com/blog/category/canadian-student-loans/
9 *Ibid.*
10 *Wikipedia, the Free Encyclopedia*, "Student loans in Canada," as posted 4 March 2017, at https://en.wikipedia.org/wiki/Student_loans_in_Canada, 2.
11 Statistics Canada, *Government Funding and Tuition as a Share of University Operating Revenue*, 1978-2008, as posted on Worthwhile Canadian Initiative: A mainly Canadian economics blog, 12 December 2011, http://Worthwhile.typepad.com/worthwhile_canadian_initi/2011/02/ca ...
12 *Wikipedia, Loc. Cit.*, "Student loans in Canada."
13 Amy Goodman, "Who is getting rich off the $1.3 trillion student debt crisis?" 29 June 2016, *Truthout.org*, as posted online at www.truth-out.org/news/item/36631-who-is-getting-rich-off-the- ..., 1.
14 Elizabeth Barrett Browning, "To George Sand: a desire."
15 List taken from the author's previous book: Thomas F. Pawlick, *Debt Sentence: How Canada's student loan system is failing young people and the country*, (Bradenton, Fla.: Booklocker.com, Inc., 2012), 10-17.
16 Charles Dickens, *A Christmas Carol* (Mineola, N.Y.: Dover Publications Inc., 1991), 5-6.
17 Jerry Kloby and Jill Schennum, "Next steps in the fight for public higher education," *Truthout.org*, 6 January 2017, as posted online at www.truth-out.org/news/item/39005-next-steps-in-the-fight-for-p ..., 2.
18 Pawlick, *Op. Cit.*
19 CBC News, "Debt woes weigh on future doctors' plans," 28 September 2011, as posted online at http://www.cbc.ca/news/health/story/2011/09/28/physician-survey-debt ...
20 Tracey Tyler, "Access to justice a basic right," *The Toronto Star*, 12 August 2007, as posted online at www.thestar.com/printarticle/245548
21 *Ibid.*
22 Robert Todd, "Legal Aid: a system in peril," *The Canadian Lawyer*, October 2010, as posted online at www.canadianlawyermag.com/Legal-aid-a-system-in-peril.html? ...
23 Canadian Bar Association, "Response to the Provost Study of Accessibility and Career Choice in the University of Toronto Faculty of Law," April 2003, 15.
24 Laura Beeston, "Nearly 40 per cent of Canadian post-secondary students experience 'food insecurity:' study," *The Toronto Star*, 2 November 2016, as posted online at www.thestar.com/news/gta/2016/11/02/nearly-40-per-cent-of-c ..., 1.
25 25 *Wikipedia, the Free Encyclopedia*, "For-profit higher education in the United States," as posted online 28 February 2017, at http://en.wikipedia.org/wiki/For-profit-higher-education-in-the-Uni ..., 3-4.
26 *Ibid.*
27 *Ibid.*
28 *Op. Cit.*, 6.
29 *Ibid.*
30 *Op. Cit.*, 7.
31 *Ibid.*

32 *Ibid.*
33 *Op. Cit.*, 1.
34 *Wikipedia, the Free Encyclopedia*, "Everest College," as posted online 28 February 2017, at https://en.wikipedia.org/wiki/Everest_College, 1& 2.
35 *Op. Cit.*, *Wikipedia*, "For-profit higher education," 5.
36 *Op. Cit.*, *Wikipedia*, "Everest College," 1.
37 Julia McKay, "Some students in line for free tuition," Kingston *Whig-Standard*, 18 January 2017, p.1.
38 *Wikipedia, the Free Encyclopedia*, "Millenials," as posted online 1 November 2016 at https://en.wikipedia.org/wiki/Millenials, 1.
39 Caelainn Barr and Shiv Malik, "Revealed: the 30-year economic betrayal dragging down Generation Y's income," 7 March 2016, *The Guardian*, as posted online at www.theguardian.com/world/2016/mar/07/revealed-30-year-ec ... 1.
40 May Warren, "Precarious work just as rampant in York Region as Toronto, report says," 13 April 2016, *The Toronto Star*, as posted online at www.thestar.com/news/gta/2016/04/13/precarious-work-just-as ..., 1.
41 *Wikipedia, the Free Encyclopedia*, "Outsourcing," as posted online 6 March 2017 at https://en.wikipedia.org/wiki/Offshoring, 2.
42 *Loc. Cit.*
43 *Op. Cit.*, 3.
44 Douglas A. McIntyre and Samuel Weigley, "States that have lost the most jobs to China," 16 September 2012, *NBC News*, as posted online 6 February 2017, at www.nbcnews.com/business/states-have-lost-most-jobs-china- ..., 1.
45 Alex Lach, "Five facts about overseas outsourcing," 9 July 2012, *Center for American Progress*, as posted online 6 February 2017 at www.americanprogress.org/issues/economy/news/2012/07/09/ ..., 2.
46 *Loc. Cit.*
47 *Ibid.*
48 Jenny Cosgrave, "Europe set to lose 1.9 million jobs, as business moves offshore: report," 20 August 2013, *ETCNBC*, as posted online 6 March 2017 at http://www.cnbc.com/id/100975246, 1.
49 *Wikipedia, the Free Encyclopedia*, "Technological unemployment," as posted online 6 March 2017 at https://en.wikipedia.org/wiki/Technological_unemployment, 2.
50 *Wikipedia, the Free Encyclopedia*, ""Financial crisis of 2007-2008," as posted online 4 January 2017 at https://en.wikipedia.org/wiki/Financial_crisis_of_2007-2008, 6.
51 *Loc. Cit.*
52 *Op. Cit.*, 7.
53 *Wikipedia, the Free Encyclopedia,* "Wells Fargo," as posted online at https://en.wikipedia.org/wiki/Wells Fargo, 28 February 2017, 1.
54 Jim Hightower, "The ethical rot of Wells Fargo, from the top down," *Buzzflash at Truthout*, 13 October 2016, as posted online at www.truthout.org/buzzflash/commentary/the-ethical-rot-of-wells ..., 14 October 2016, 1.
55 Renae Merle, "Wells Fargo CEO steps down in wake of sham accounts scandal," *The Washington Post*, 12 October 2016, as posted online at www.washingtonpost.com/news/business/wp/2016/10/12/wells ..., 3 February 2017.
56 Matt Egan, "Wells Fargo scandal spreads to Prudential insurance," *CNNMoney*, 12 December 2016, as posted online at http:// money.cnn.com/2016/12/12/investing/wells-fargo-insurance-sca ..., 3 February 2017, 1.
57 *Loc. Cit.*
58 *Wikipedia, the Free Encyclopedia*, "Goldman Sachs," as posted online at https://en.wikipedia.org/wiki/Goldman_Sachs, 2 March 2017, 13.
59 Adam Gabbatt, "Former Goldman Sachs trader found guilty of mortgage fraud,"

Chapter Four

EARTH, AIR, FIRE AND WATER

"The concept of global warming was created by and for the Chinese in order to make U.S. manufacturing non-competitive."
—Tweet by Donald Trump, 6 November 2012.

"Obama's talking about all of this with the global warming and ... a lot of it's a hoax. It's a hoax."
—Speech by Donald Trump at Hilton Head, S.C., 30 December 2015

Aunt Helen was a lovely woman, sweet tempered, gentle, good-hearted. She had a puckish sense of humour, and when one of her subtler jokes finally dawned on someone, her smile and laughing eyes lit up the room. Her eyes and her smile were the best points of her considerable beauty. Intelligent and unusually well-read, she also had a practical side, expressed through her studies at *Cordon Bleu*, which made her a superb cook. Being around her, and Uncle Andy, a Second World War combat veteran and my favourite uncle, was always a joy.

Unfortunately, both were heavy smokers, Andy of pipes and Helen of cigarettes, and Helen died far too young, of cancer. It was an awful, ugly death, slow and agonizing. Ma went out to the coast to help when she reached the terminal stage, and was shocked by what she saw. So was Andy, who suffered a heart attack not long after Helen died.

In clinical terms, the usual progress of the disease that killed Helen is straightforward:

> *Early symptoms may include lingering or worsening cough; coughing up phlegm or blood; chest pain that worsens when you breathe deeply, laugh or cough; hoarseness; shortness of breath, wheezing, weakness*

of experiment and direct observation of physical facts. Empiricism was the key to it all. He believed that the concepts people came to hold were determined by their perceptions of the world around them, and so he strove above all for accuracy of perception.

For example, in his book *Generation of Animals*, he recounts breaking open fertilized chicken eggs at different time intervals after laying, to understand the sequence of appearance of the various organs. He was a close observer as well of sea life and employed dissection to describe and understand the internal organs of fish and octopi.

Later investigators built on his ideas. The Iraqi Ibn al-Haytham (965-1039), for example, pioneered the idea of using structured experiments to answer specific questions, and of requiring that experimental results, in order to be valid, ought to be repeatable by other observers who try the same experiments.

Over the years, pioneering individuals, including the philosopher William of Ockham (1285-1347), early astronomers Johannes Kepler (1571-1630) and Galileo Galilei (1564-1642), and dozens of others, such as Charles Darwin in the 19th and Albert Einstein in the early 20th century, expanded and refined the ideas of their predecessors until, at last, today's structure was in place.

In an eight-step nutshell (with thanks to *Wikipedia*[15]), it goes like this:

1) Define a question you hope your work will answer;
2) Gather information, including statistical data and/or direct physical observations;
3) Form an explanatory hypothesis;
4) Test the hypothesis by performing an experiment and/or highlighting data in a reproducible way;
5) Analyze the data, logically and where required mathematically;
6) Interpret the data and draw conclusions that confirm the hypothesis and serve as a starting point for new hypotheses;
7) Publish the results, preferably in a reputable, peer-reviewed journal not dependent on any one commercial enterprise or sector for its funding;
8) Retest/repeat the experiment (usually done by other scientists).

The requirement of *peer review* is essential. Generally, reputable journals like *Nature,* or *Science*, have expert panels on-call to review work submitted for publication. Before a line goes into print, these experts, mostly specialists in the same discipline as the would-be contributor, read the data and give their opinion as to its quality and relevance. Some may actually perform the contributor's experiment(s) themselves to verify that the work is reliable.

Only if the review panel approves is the article published. This rule applies normally to reports published in the main, or "news" section of a journal, in which factual material dominates. The opinion, or editorial section, where editors or guest columnists contribute personal opinion pieces, clearly labelled as opinion, are not always peer-reviewed.

Once published, other scientists who read the journal are able to criticize, sending in letters to the editor or even providing contrary data they believe could invalidate the contributor's conclusions. Critics often suggest other, simpler explanations for the same phenomena, according to the principle first proposed by William of Ockham, known as Ockham's Razor. William, a Franciscan monk, wrote it in Latin, *entia non sunt multiplicanda praeter necessitatem,* or "entities should not be multiplied beyond necessity."[16] In other words, the simplest explanation that fits the available data, making the fewest assumptions and containing the least number of hypothetical points, is preferable over longer, more complicated ones.

In the vernacular, cut out the bullshit (with a sharp razor).

Some of the more respected science journals are read by hundreds of thousands of well-informed readers all over the world, many of whom have spent between six and 10 years or more at university, studying their subjects. Those with graduate degrees had to write theses of their own, in the form of masters or doctoral dissertations, and then defend them before a live, expert panel at their university, before getting their degrees. This author knows from personal experience what that gruelling process is like. After that they may also have spent years in the field, doing research similar to that of the would-be journal author. The original experimenter is thus very much on the hot-seat, and must defend his or her ideas with energy.

If there is sufficient information or scope in one or several such published reports to explain a significant portion of a discipline, the explanation may be elevated to the status of a theory. The term *theory* "represents a hypothesis (or group of related hypotheses) which *has been confirmed through repeated testing, almost always conducted over a span of many years*. Generally, a theory is an explanation for a set of related phenomena."[17]

Einstein's theories of Special and General Relativity, accepted today by probably 99.9 per cent of all physicists as the correct explanation for what he observed, are an example. If Einstein was wrong, then all those atom and hydrogen bombs various governments have exploded should never have made a bang. Nor should the reactors at Three Mile Island, Chernobyl or Fukushima have caused alarm when they melted down.

Thus, calling a scientific statement "merely theory" is to ignore or deprecate a serious, significant body of detailed work, whose validity has stood the test of time and which has been accepted as accurate by thousands, sometimes millions, of reputable scientists and technicians.

Statistics, lies and law

Not all scientific reports deal with experiments in a laboratory, using test tubes, Bunsen burners and other such apparatus. Some, particularly in the medical field of epidemiology, deal with statistical evidence. The studies supervised by the Polish Academy of Sciences and the Phagoburn study, mentioned in Chapter One, are examples. To understand them completely requires a good head for math, which means at least passing grades in calculus. For the general reader, however, a few points should suffice.

Essential to any numerical study are the control group, sample size, absence of bias, and the concept of statistical significance. For instance, if you want to assess the ability of HeadKrusher brand helmets to prevent concussions among Junior League hockey players, you must compare at least two groups of players, one wearing HeadKrushers and the other wearing another brand or brands. The non-HeadKrusher group is the control.

You must also survey a large enough number of players, your sample size, over a long enough time period, such as a full hockey season, for the results to have any meaning. Comparing two groups of 10 players each for one week would result in nothing helpful. Perhaps none of them would get concussed, or those who did may by sheer chance have been cross-checked into the boards twice as many times as the others, introducing a random variable that would invalidate the results. There are a number of mathematical formulae for setting sample sizes, depending on the type of survey being conducted, but they all boil down to one simple rule: the Bigger and Longer the Better.

Yes, it's true ...

It's important as well to avoid bias in any study, meaning that those involved in and those conducting the study should not have an axe to grind, such as being salespersons for HeadKrusher Helmets Inc. Blind tests, where the subjects don't know whether they are in the control group or are the actual test subjects, and double-blind tests, where even the conductors of the survey don't know which group is which, supposedly correct for bias.

There are also fairly strict, and complex, formulas for establishing the statistical significance of a test result. In fact, "books have been filled with the collected criticism of significance testing."[18] The most frequently used rule is probably (so to speak) the "five per cent rule," originally proposed in 1925. Under it, data are deemed statistically significant only if the probability "p" of a phenomenon being random is *less* than one chance in 20, or five per cent. Thus, if all people holding nails normally hit their thumbs with a hammer once in 20 swings, or five per cent of the time, it's not significant (pain is not the criterion here). But if a big enough group of people consistently hit their thumbs only one time in every 40 swings, with only a two per cent probability of that's being random, it is significant. Some other factor should be suspected, such as all the hammer swingers being skilled professional carpenters.

This is a confusing rule, since it is defined negatively, but statistics are like that, which is why some of us only took the course for one semester. Statistics can also be taken out of context, or twisted to suit a debater's purposes, which gave rise to the saying, popularized by Mark Twain, "lies, damned lies and statistics." Be on guard when they pop up.

Scientists and mathematicians love to dicker over fine points, of course, and there are times when getting them to agree on anything, including statistics, seems like herding cats. For practical purposes, however, the dickering has to stop somewhere, and a court of law is one place where such a line is drawn. In the U.S., it was drawn fairly definitively by the Supreme Court, in Washington, in the 1993 *Daubert vs. Merrell Dow Pharmaceuticals, Inc.* decision.

The Justices quoted an *Amici Curiae* (Friends of the Court) brief submitted to the court by the American Association for the Advancement of Science and the National Academy of Sciences, which noted that "Science is not an encyclopaedic body of knowledge about the universe. Instead, it represents a process for proposing and refining theoretical explanations about the world that are subject to further testing and refinement."[19] Accordingly, the Court said, "a key question to be answered in determining whether a theory or technique is scientific knowledge that will assist the trier of fact will be whether it can be (and has been) tested." Judges must decide the "evidential reliability" of scientific conclusions based on the *methods used to test them.*[20]

As Caltech professor David Goodstein writes in his summary of the case:

> *In particular, the methods should be judged by the following four criteria:*
>
> 1) *The theoretical underpinnings of the methods must yield testable predictions by means of which the theory could be falsified;*
> 2) *The methods should preferably be published in a peer-reviewed journal;*
> 3) *There should be a known rate of error that can be used in evaluating the results;*
> 4) *The methods should be generally accepted within the relevant scientific community.*[21]

So, tests, with a known rate of error, using generally accepted methods and published in peer reviewed journals: a pretty good rule of thumb, but not, obviously, foolproof. Nothing done by mere mortals is perfect. And this is where the all important **Precautionary Principle** comes into play.

Not even the most authoritative body of scientific experts and/or jurists can be right all of the time, but a line must still be drawn somewhere in order for life, and practical affairs to go on. To avoid recklessness, or overly risky action that could harm the innocent, science and law fall back on the principle established long ago by the ancient Greek medical theorist Hippocrates, in his book *Epidemics*: "The physician must ... have two special objects in view with regard to disease, namely to do good or *to do no harm*."[22] Later refined to the Latin saying "*primum non nocere* (first, do no harm)," it became the touchstone of medicine.

In the 20th century, the principle was expanded beyond medicine to apply more generally throughout science, as well as in law. A basic axiom in risk management today, it states that "if an action or policy has a suspected risk of causing harm to the public, or to the environment, in the absence of scientific consensus (that the action or policy is not harmful) the burden of proof that it is *not* harmful falls on those taking the action."[23] Or, if you want us to let you do it, prove it won't hurt. It has become a statutory requirement in law in numerous jurisdictions, including the European Union, and was endorsed worldwide in 1982 by the United Nations General Assembly.

How do all of these methods, rules and principles of science work out, where things like cancer or climate change are concerned? The Tobacco Wars provide a good illustration.

A rare oddity

Until the late 19th century, lung cancer was "a very rare disease, so rare that doctors took special notice when confronted with a case, thinking it a once-in-a-lifetime oddity."[24] According to a detailed retrospective in the *BMJ Journals* (previously the *British Medical Journal*), "Lung cancer was not even recognized medically until the 18th century, and

In a public *Frank Statement to Cigarette Smokers* in 1954, the industry "set the tone"—and pretty much gave its own game away—claiming that:

> Distinguished authorities point out:
>
> 1) That medical research of recent years indicates many possible causes of lung cancer;
> 2) That there is no agreement among the authorities regarding what the cause is;
> 3) That there is no proof that cigarette smoking is one of the causes;
> 4) That statistics purporting to link smoking with the disease could apply with equal force to any one of many other aspects of modern life. Indeed the validity of the statistics themselves are questioned by numerous scientists.[50]

As a Philip Morris executive later quipped: "Anything can be considered harmful. Apple sauce is harmful if you get too much of it."[51]

Trying to convince audiences that each of the four points was true, or at least arguable (even if they weren't), became the goal. As Hill and Knowlton staffer Carl Thomson advised the industry magazine *Tobacco and Health Research* in a 1968 letter: "The most important type of story is that which casts doubt in the cause and effect theory of disease and smoking. Eye-grabbing headlines were needed and should strongly call out the point—Controversy! Contradiction! Other Factors! Unknowns!"[52]

The creation of doubt, as a goal in itself, was echoed roughly a year later in what became an infamous memo by an unknown Brown and Williamson executive, urging that advertising be used to "counter the anti-cigarette forces." The memo states: "Doubt is our product,[53] since it is the best means of competing with the 'body of fact' [linking smoking with disease] that exists in the mind of the general public. It is also the means of establishing a controversy ... If we are successful in establishing a controversy at the public level, there is an opportunity to put across the real facts about smoking and health."[54]

A companion Brown and Williamson memo, also dated 1969, lists the ad campaign's objectives:

> *Objective No. 1: To set aside in the minds of millions the false conviction that cigarette smoking causes lung cancer and other diseases; a conviction based on fanatical assumptions, fallacious rumours, unsupported claims and the unscientific statements and conjectures of publicity-seeking opportunists.*
> *Objective No. 2: To lift the cigarette from the cancer identification as quickly as possible and restore it to its proper place of dignity and acceptance in the minds of men and women in the marketplace of American free enterprise.*
> *Objective No. 3: To expose the incredible, unprecedented and nefarious attack against the cigarette, constituting the greatest libel and slander ever perpetrated against any product in the history of free enterprise.*
> *Objective No. 4: To unveil the insidious and developing pattern of attack against the American free enterprise system, a sinister formula that is slowly eroding American business with the cigarette obviously selected as one of the trial targets.*[55]

In the 1980s, during the second-hand smoke debates, "UK tobacco researchers commented on how Philip Morris ... was piloting a 'global strategy' to deny the reality of second-hand smoke hazards, spending vast sums of money 'to keep the controversy alive.'"[56] Tools in general use by the industry included:

- Delaying and discrediting legitimate research;
- Promoting "good" epidemiology, and attacking so-called "junk science" (a term popularized by industry lobbyist Steven Milloy);
- Creating outlets for research favourable to industry. It was found that in 1989, the tobacco industry set up a supposed scientific journal, *Indoor and Built Environment*, "which published a large amount of material on passive smoking, much of which was "industry-positive." It was later found that the executive and editorial boards of the publication were "dominated by paid tobacco-industry consultants."[57]

1994 paper, he argued that "there is no good scientific evidence that passive inhalation [of tobacco smoke] is truly dangerous under normal circumstances."

Singer has argued in books and articles that there is no evidence that global warming is attributable to human-produced CO_2, and claims that humanity "would benefit"[65] if temperatures rise. Seitz "was the highest ranking scientist among a band of doubters who, in the early 1990s, resolutely disputed suggestions that global warming was a serious threat." He was criticized by the National Academy of Sciences (NAS) for having written an article imitating the style and format of the *Proceedings* of the NAS, including a spurious date of publication and volume number, when no such paper had actually been published in the journal.[66]

As for Myron Ebell, who unlike Singer and Seitz is not a scientist, he has become something of a bogey man among environmentalists for his efforts to discredit the science of global warming, first as a spokesman for right wing think tanks, and later in government.[67] Many others, including Milloy, Horner, Richardson and Enstrom, are working in step with Ebell (see below), walking the path Big Tobacco pioneered.

But what, exactly, are they attacking? What does science say about global warming and climate change?

How climate science works

Unlike weather prediction, which deals with short-term (up to a week) and purely local phenomena, climate science looks at long-term trends, observed over significantly large regions or even the entire planet. It is firmly based on physics and geography, two very solid pillars of science, and on direct observation. Its practitioners have reported two important findings: a) average global temperatures on the earth's surface are rising, and have been rising for some time; b) it is "extremely likely (meaning 95 per cent probability or higher)"[68] that the rise cannot be explained by any factor other than human activity.

This is not new, nor controversial, at least among scientists, but has been known and all but unanimously accepted by them for years. It is not hypothesis or even theory, but simple, observed fact. What the effects of these two facts might be on the environment, and what consequences they might impose on human society, may be debated. But the facts themselves are not being argued—except, of course, by the denial industry.

What is the evidence for climate warming, and for human activity as its leading cause? And what do the deniers say about it?

The first proof that global temperatures are warming is actual thermometer readings, taken over roughly the past 150 years, or since thermometers have been available (the first mercury thermometer was invented in 1714) and placed in sufficient numbers on land and on ships and buoys at sea. A bit of math is involved, to correct for instrumentation and placement in different regions, but the results are clear. Since about 1860, the record shows:

1) A roughly constant temperature in the late 19th century;
2) a rise in the early 20th century;
3) a levelling off or slight decline in the mid-20th century;
4) a steep rise in the final decades of the 20th century, continuing into the 21st century.[69]

According to the U.S. government National Centers for Environmental Information, of the National Oceanic and Atmospheric Administration (NOAA), global average surface temperature "shows an increase of approximately 1.4 degrees Fahrenheit since the early 20th century."[70]

[Recent claims of a seeming "pause" or "hiatus" in rising temperatures during the first decade of the 21st century, noted by the Intergovernmental Panel on Climate Change (IPCC), have turned out to be in error, as a University of Bristol study[71] and a 2015 report in *Science*[72] have shown.]

Evidence for the planetary rise in temperature is not limited to open-air thermometer readings, but also includes measurements of sea levels, upper ocean water heat increases, the retreat of average annual snow cover in the northern hemisphere, and shrinking worldwide glacier volumes.[73] A four-decade increase in the Climate Extremes Index is also indicative of such a warming trend.[74] Plant and animal species range changes also show that northern hemisphere species are moving northward at a rate of six kilometres per decade, as warmer conditions expand.

Putting this current warming trend into long-term geological context (the paleo-climate) requires going back thousands or even millions of years before humans began using thermometers. The record here is more complex, but based on several kinds of reliable experimental observation, such as:

> **Annual tree rings** (observed in horizontal sections of tree trunks, or fossilized trunks), whose width and density vary with temperature and length of the growing season;
> **Coral reef annual layers**, showing growth in calcium carbonate deposits that correlate with tropical ocean temperature;
> **Lake sediments**, whose thickness shows the rate of snowmelt feeding streams, and whose pollen content shows plant species living at the time of sediment deposit;
> **Isotope ratios**, taken from ice core borings, specifically ratios of oxygen-16 to oxygen-18 (heat evaporates oxygen-18, so it is measurably less common in warmer climates), as well as hydrogen isotope ratios.

Such evidence, particularly ice core samplings from Greenland and Antarctica, using isotope ratios to estimate temperatures over roughly a million-year span, "show a cyclic pattern of brief (10,000 to 20,000-year) warm spells called *interglacials*, separated by longer cold spells (*ice ages*). For the past half million years, this cycle has repeated on roughly 100,000-year intervals," resulting from "subtle changes in earth's orbit and tilt."[75]

The recent temperature spike departs from this pattern. In fact, it runs counter to what one should expect from changes in the earth's tilt or orbit. "During the last few thousand years, this phenomenon [orbit and tilt] contributed to a slow cooling trend at high latitudes of the Northern Hemisphere during summer, a trend that **was reversed** by greenhouse gas induced warming during the 20th century."[76]

So, the earth, especially the northern region where there is more land area, is observed to be warming, at a time in observed geological history when at least the Northern Hemisphere ought to be cooling. Earth's current climate is now "the warmest in at least the past millennium."[77]

Other than changes in planetary tilt or orbit, natural causes of warming or cooling also include variations in light intensity emitted by the sun, and the presence in the atmosphere of dust particulates and/or gases emitted by volcanic eruptions.

"Since 1978, solar irradiance has been measured by orbiting [space] satellites. These measurements indicate that the sun's radiative output has not increased since then, so the warming that occurred in the past 40 years cannot be attributed to an increase in solar energy reaching the earth."[78]

As for volcanic activity, its overall impact on climate tends more toward cooling than warming. The spectacular 1883 eruption of Mt. Krakatoa in the Dutch East Indies, for example, was followed by a drop in Northern Hemisphere summer temperatures of 2.2 degrees F. This was due to the injection of "an unusually large amount of sulphur dioxide (SO_2) gas high into the stratosphere, which was subsequently transported by high-level winds all over the planet. This led to a global increase in sulphuric acid (H_2SO_4) concentration in high-level cirrus clouds. The resulting increase in cloud reflectivity (or albedo) would reflect more incoming light from the sun than usual, and cool the entire planet until the suspended sulphur fell to the ground as acid precipitation."[79]

In some cases, volcanic eruptions can also cause short-term, localized warming effects, but the overall cooling pattern, or "volcanic winter" is paramount. In any case, such phenomena are relatively short-lived. The chief weather effects of Krakatoa lasted only about five years, while that of other, more recent volcanoes has been on the order of one to two years. This could not explain the 40-year global rise in average surface temperature thus far recorded.

What does explain it, according to a long and still-lengthening list of published scientific reports, is a rise in the percentage of so-called greenhouse gases in the atmosphere. That these have been increasing, and in sufficient amounts to cause the thus-far-recorded warming, is as much a matter of factual observation as the rise in temperatures itself.

In the hothouse

The terms "greenhouse gas" or "greenhouse effect" refer to the tendency of certain gases to block or reflect infrared radiation. Energy from the sun, in the form of sunlight, reaches the earth at an average rate of 240 watts per square metre, but a portion of it is then re-radiated back into space in the form of infrared radiation. A stable climate results from a balance between incoming solar energy and the heat that our planet radiates back to space.

Most of earth's atmosphere, which is made up chiefly of nitrogen and oxygen, "is transparent both to incoming sunlight and to outgoing infrared," but some atmospheric gases "are transparent to sunlight but considerably opaque to infrared."[80] These include water vapour (H_2O), methane (CH_4) and carbon dioxide (CO_2). They let the sun's energy in, but not out again, something like the glass covering a greenhouse. (The name is actually somewhat of a misnomer, as Prof. Richard Wolfson explains, "because the glass in a greenhouse blocks predominantly the bulk motion of heated air out of the greenhouse, not infrared radiation.")[81]

Water vapour is the most prevalent greenhouse gas, but not the "prime mover" where global warming is concerned. Although a potent blocker of infrared radiation, the "typical water molecule remains in the atmosphere only about a week before it is removed"[82] as rain or other precipitation. And something else has to get it up into the atmosphere first, before it can do any blocking. That is, something else has to do the initial warming, in order to evaporate the water. Rather than causing the initial temperature rise, water vapour only amplifies it after something else has launched the process. That something else is CO_2.

In system terms, rising CO_2 is an input, which causes warming as its output, which then creates a feedback mechanism in the form of rising water vapour, which establishes a positive feedback loop that amplifies the heat.

Methane is also a factor, but there is considerably less of it in the atmospheric mix, and "its lifetime in the atmosphere is only about 12 years before chemical reactions remove it permanently."[83] In contrast, any carbon that is added to the overall global carbon system is not removed permanently but continues cycling through the atmosphere, biosphere (via plants) and oceans "for hundreds to thousands of years," so that "the level of atmospheric carbon dioxide goes up and stays up"[84]—and continues evaporating water.

Normally, the amount of such gases in our atmosphere keeps average surface temperatures comfortable for present-day life forms—namely about 33 degrees Celsius or 60 Fahrenheit higher than they would be without our atmosphere. But it doesn't take a huge deviation from this normal to make things difficult for living creatures who are adapted to it. For example, the difference between a relatively warm interglacial period and that of the most recent ice age, when the present site of Montreal was covered by a thick layer of ice, is only about 6 degrees Celsius.[85]

The amount of greenhouse gases in our atmosphere—particularly CO_2—has risen just enough over recent years to account for the current observed average global warming. No other known cause corresponds so closely. What is the evidence for this?

Direct measurements of atmospheric gases, usually expressed in parts per million (ppm), are taken regularly at hundreds of weather stations in more than 60 countries around the globe.[86] The best known of these is probably the station atop Mt. Mauna Loa, in Hawaii, chosen as an observation site because, located far from any continent its air "is a good average for the central Pacific. Being high, it is above the inversion layer where most of the local [weather] effects are present ... The contamination from local volcanic sources is ... removed from the background data."[87]

The first "reproducibly accurate"[88] measurements of atmospheric CO_2 with manmade instruments began in the 1950s, with Mauna Loa's measurements beginning in 1956-58. For concentrations before such instrumentation was available, scientists use the same sort of ice core samples employed to measure temperatures from earlier geologic eras. Ice cores from the Antarctic or Greenland ice sheets, for example, contain so-called "fossil" air bubbles indicating that concentrations of CO_2 "were about 260-280 ppm by volume immediately before industrial emissions began and did not vary much from this level during the preceding 10,000 years [of our present interglacial period]. The longest ice core record comes from East Antarctica, where ice has been sampled to an age of 800,000 years."[89]

Readings from manmade instruments since the 1950s, plus those from the ice core samples for earlier years, plus less precise analyses of boron and carbon isotope ratios in some marine sediments, show that:

a) atmospheric CO_2, previously stable for thousands of years, has been rising since the beginnings of the industrial revolution, from approximately 280 ppm in the mid-18th century to 403 ppm in 2016, with a more rapid rise from around 300 ppm in the 1950s to today's 400-plus ppm levels;[90]
b) expressed in graphs, the pattern of these rises in atmospheric CO_2 corresponds remarkably closely to the recorded rise in global temperature. In fact, rises and drops in atmospheric CO_2 have shown a consistently close correlation to global temperature going back as far as 400,000 years before the present;[91]
c) the rate of rise in atmospheric CO_2 also corresponds closely to the recorded rise in human-origin global CO_2 emissions since the Industrial Revolution, as opposed to any natural cause. A particularly close correlation is recorded on a graph from the Scripps Institute showing the rise of cumulative industrial CO_2 emissions in tandem with monthly average atmospheric CO_2 concentrations in ppm from the 1950s through 2012.[92]

As long as there are no changes in planetary tilt or orbit, or in light emissions from the sun, "natural CO_2 emissions are balanced by natural absorptions,"[93] and thus the overall amount of CO_2 in the atmosphere should remain fairly stable. The rises noted are thus likely not natural in origin.

The two leading sources of human produced CO_2 are the burning of fossil fuels such as coal, oil and natural gas, and forest destruction, which releases the CO_2 stored in trees and forest plants. Since fossil fuels are commercial products, recorded in financial transactions, and logging and forest clearing are also generally recorded, we have a fairly accurate record of both. Fossil fuel combustion "results in about 7 gigatonnes per year of carbon being emitted to the atmosphere in the form of CO_2," while "roughly 2 Gt of carbon per year results from deforestation"[94] as well as agricultural practices. This, also, is a matter of observation rather than speculation.

This is because natural sources of CO_2 are normally balanced between emissions (caused by such things as the respiration of animals and soil organisms, which exhale CO_2, by volcanoes, by ocean/atmosphere exchange, etc.) and absorption (via plants, which take in CO_2 and give off oxygen, by ocean/atmosphere exchange and soil activity). "The amount of carbon dioxide produced by natural sources is offset by natural carbon sinks [absorbers of carbon] and has been for thousands of years. Before the influence of humans, carbon dioxide levels were quite steady because of this natural balance."[95] As noted in (a) above, this began to change only after the human industrial revolution.

Another very strong indicator of their human origin is the ratio of various carbon isotopes in the CO_2 being added to the atmosphere. Carbon-14 is an unstable radioactive isotope with a half-life of 6,000 years, and thus is absent from fossil fuels formed millions of years ago. Oil, coal and other such sources contain virtually no C-14. They also contain relatively little carbon-13, since they are derived mostly from plants, which take up less C-13 than they do carbon-12. Any CO_2 emitted by burning fossil fuels should thus contain no C-14, little C-13 and instead be made up chiefly of C-12.

This is, in fact, the case. "Studies of tree rings have shown that the proportion of carbon-14 in the atmosphere dropped by about two per cent between 1850 and 1954 [when atomic bomb testing skewed subsequent readings]." At the same time, "the ratio of carbon-13 to carbon-12 in the atmosphere and ocean surface waters is steadily falling, showing that more carbon-12 is entering the atmosphere."[96]

As Prof. Richard Wolfson, of Middlebury College, explains:

> Fossil carbon has been buried for hundreds of millions of years. Therefore, it is depleted in the radioactive isotope carbon-14. Furthermore, living things take up stable carbon-12 more readily than they

> do stable carbon-13, and living things are the origin of the fossil fuels. Therefore, fossil carbon is also depleted in C-13. Examination of the isotopic composition of atmospheric carbon shows that it is becoming increasingly depleted in C-13 and C-14, indicating the fossil origin of the added carbon.[97]

An article in *New Scientist* makes the same point: "Studies of tree rings have shown that the proportion of C-14 in the atmosphere dropped by about two per cent between 1850 and 1954. After this time, atmospheric nuclear bomb tests wrecked this method by releasing large amounts of carbon-14." However, "fossil fuels also contain less carbon-13 than carbon-12, compared with the atmosphere, because the fuels derive from plants which preferentially take up the more common carbon-12. The ratio of carbon-13 to carbon-12 in the atmosphere and ocean surface waters is steadily falling, showing that more carbon-12 is entering the atmosphere."[98]

There are other indicators as well, including decreases in atmospheric oxygen, increasing acidification of the oceans, and other phenomena. Combined, they show an unmistakable human "footprint."

The vast majority of qualified scientists and scientific organizations of international standing agree that "warming of the climate is unequivocal" and "most of the global warming since the mid-20th century is very likely due to human activities." They also agree that "overall, net effects are more likely to be strongly negative with larger and more rapid warming" and that "the resilience of many ecosystems is likely to be exceeded this century by an unprecedented combination of climate change, associated disturbances (e.g. flooding, drought, wildfire, insects, ocean acidification) and other global change drivers ... **No scientific body of national or international standing maintains a formal opinion dissenting from any of these main points.**"[99]

A partial list

A partial list of synthesis reports and public statements by scientific organizations backing these views, given by the *Wikipedia* under "Scientific opinion on climate change"[100] includes:

First and foremost, the **Intergovernmental Panel on Climate Change (IPCC)**, set up in 1988 by the World Meteorological Organization (WMO) and the United Nations Environment Program (UNEP) following pressure from the conservative Republican U.S. government of Ronald Reagan. Reagan was worried that the IPCC's predecessor scientific body, the Advisory Group on Greenhouse Gases, was inadequate to its task and allowed "unrestrained influence from independent scientists."[101] The new IPCC was organized to assure its reports would be reviewed by all sponsoring governments (roughly 194 of them[102]) and that its summaries would be "subject to line-by-line approval by delegates from all participating governments."[103]

The IPCC's periodic **Assessment Reports** are based on a comprehensive review of the published scientific literature, some 90 per cent of which is peer-reviewed.[104] "Thousands of scientists contribute (on a voluntary basis, without payment from the IPCC) to writing and reviewing the reports,"[105] whose Summary for Policymakers is, as noted above, subject to line-by-line approval by delegates from all participating countries. IPCC drafts are said to "undergo more scrutiny than any other documents in the history of science."[106]

Synthesis reports backing the main points have also been issued by the **U.S. Global Change Research Program** and the intergovernmental **Arctic Council** and non-governmental **International Arctic Science Committee**.

Scientific bodies of national and international standing which have issued individual formal statements concurring with the points mentioned include "34 national science academies, three regional academies, and both the international **InterAcademy Council** and **International Council of Academies of Engineering and Technological Sciences**. Numerous academies have also signed on to joint, as opposed to individual statements, including **the science academies of Australia, Belgium, Brazil, Canada, the Caribbean, China, France, Germany, India, Indonesia, Ireland, Italy, Japan, Malaysia, Mexico, New Zealand, Russia, South Africa, Sweden, Turkey, the United Kingdom and the United States.**

Also signing joint statements were the **Network of African Science Academies**, and the science academies of the **G8 + 5 nations**.

Additional backers of the main points, in the form of statements, are **the American Association for the Advancement of Science (the world's largest general scientific society), the Federation of Australian Scientific and Technological Societies, the U.S. National Research Council, the Royal Society of New Zealand, the Royal Society of the United Kingdom, the African Academy of Sciences, the European Academy of Sciences and Arts, the European Science Foundation, the American Geophysical Union, American Chemical Society, American Institute of Physics, American Physical Society, Australian Institute of Physics, European Physical Society, American Academy of Agronomy, Crop Science Society of America, Soil Science Society of American, European Federation of Geologists, European Geosciences Union, Geological Society of America, Geological Society of London, International Union of Geodesy and Geophysics, National Association of Geoscience Teachers, American Meteorological Society, the Australian Meteorological and Oceanographic Society, the Canadian Foundation for Climate and Atmospheric Sciences, Canadian Meteorological and Oceanographic Society, Royal Meteorological Society (UK), World Meteorological Organization, American Quaternary Association, International Union for Quaternary Research, American Association of Wildlife Veterinarians, American Institute of Biological Sciences, American Society for Microbiology, Australian Coral Reef Society, Institute of Biology (UK), Society of American Foresters, The Wildlife Society, American Academy of Pediatrics, American College of Preventive Medicine, American Medical Association, American Public Health Association, Australian Medical Association, World Federation of Public Health Associations, American Astronomical Society, American Statistical Association, Canadian Council of Professional Engineers, The Institute of Engineers Australia, International Association for Great Lakes Research, Institute of Professional Engineers New Zealand, the World Federation of Engineering Organizations.**

Listed as noncommittal, or expressing only partial agreement—though not opposing the points—were the **American Association of**

Petroleum Geologists, American Institute of Professional Geologists and the Canadian Federation of Earth Sciences, all three bodies with memberships heavily dependent on the fossil fuel industry for their livelihoods.

The U.S. National Aeronautics and Space Administration (NASA)—the folks who put men on the moon—also agrees that the climate is warming due to human activity. It refers visitors to its Global Climate Change site[107] to another, still longer list of world scientific organizations backing the position that "climate change has been caused by human action." The list, published by the **California Governor's Office of Planning and Research**, contains 198 scientific organization names.[108] These organizations and societies speak for, literally, tens of thousands of their scientist/members.

Besides the IPCC Assessments, and the statements of so many scientific associations and groups, there have also been numerous independent surveys of the published scientific literature, of which some of the most important are:

- **University of California science historian Naomi Oreskes' 2004 paper in Science magazine, "The Scientific Consensus on Climate Change,"** which analyzed "928 abstracts published in refereed scientific journals between 1993 and 2003, and listed in the ISI database with the key words global climate change." She found that "none of the papers disagreed with the consensus position" of the IPCC that "Earth's climate is being affected by human activities ... [and] most of the observed warming over the last 50 years is likely to have been due to the increase in greenhouse gas concentrations." She concluded that "there is a scientific consensus on the reality of anthropogenic climate change."

The paper contained a minor typographical error, dropping the word "global" from the phrase "global climate change" in one paragraph, which error was corrected in the next issue of the magazine. SDI critics, as will be described in greater detail below, pounced on this.

Gingrich's proposals to reform the Endangered Species Act, which noted that Gingrich's "soft feelings for cuddly little critters is still going to be a big problem."

Climate change deniers often adopt such a derisive, belittling or insulting tone, reminiscent of Internet trolls. Verbal debating techniques may involve talking over or drowning out opponents, and repetition of so-called "talking points," to fix certain phrases in viewer/listeners' minds. These tactics play well with the target audience identified by the now-defunct Information Council on the Environment (ICE), which in 1991 developed an anti-climate warming publicity plan similar to that of the earlier tobacco industry playbooks, geared to "reposition global warming as theory (not fact)."[127] The ICE was "created by the National Coal Association, the Western Fuels Association and Edison Electrical Institute."[128] Its campaign was aimed at:

> *"[O]lder, less-educated males from larger households who are not typically active information-seekers" and "younger, low-income women" as "good targets for radio advertisements" that would "directly attack the proponents of global warming ... through comparison of global warming to hysterical or mythical instances of gloom and doom." One print advertisement featured a cowering chicken under the headline "Who told you the earth was warming—Chicken Little?"*[129]

"He [Ebell] doesn't really know anything about science," said Gavin Schmidt, director of the Goddard Institute for Space Studies and a top earth scientist at NASA who has faced off with Ebell in the past. "He uses science like a talisman." Ebell's technique, Schmidt said, is to point toward "some little fact" and use it to extrapolate some larger irrelevant and scientifically incorrect point.[130]

Science denialists in general use a similar tactic, focused on casting doubt upon "keystone dominos," or isolated pieces of evidence that capture people's attention, like polar bears or arctic ice floes. If one piece can be made to look questionable, "audiences targeted by the communication may assume all other dominoes are toppled in a form of 'dismissal by association.'"[131] Republican U.S. Senator Jim Inhofe, from an oil producing state, used the tactic in 2015 when he brought a snowball into the senate, laughingly derided global warming, and shouted "catch this!" as he threw the ball to his audience.[132]

The method has been compared to a fictional scenario of the publishers of the English-language *Webster's Dictionary* hypothetically deciding they don't want to compete with the Larousse company, publishers of French dictionaries. They look for a way to discredit Larousse. They hire the denial industry specialists, who start putting out news releases and articles claiming that English and French are not two separate languages at all, that there is only English and this "so-called French language" is just "fake."

As proof, they point to words that are the same in both languages. "Look it up for yourself in this Larousse so-called French dictionary," they say. "For example, find the word 'international,' and you'll see it's the same word in BOTH dictionaries, spelled exactly the same way. And that's not all, look up 'gazelle.' Same word in both dictionaries, spelled just the same. And paper, in the Larousse it's misspelled 'papier,' which just shows how sloppy the Larousse people are. They don't even bother to proofread their pages! There is NO FRENCH LANGUAGE! It's all a myth, a fake. You only need our English dictionary."

If you ask why there are so many French teachers at so many universities claiming "so-called French" is a separate language, the reply is: "Of course they say that! They get paid to maintain the fiction. Their salaries depend on it."

To someone who doesn't speak both languages, and happens also to be rather a dunce, it might even make sense. And in the same way, the denial industry's PR specialists cast doubt on scientific ideas, and ultimately on science itself. The legitimacy of all science, and every scientific statement is called into question. In the end, the public loses faith in the entire scientific enterprise, which is a terrible travesty of truth.

The adage "follow the money" doesn't work consistently, either, in attempting to separate the wheat from the chaff. Due to hidden funding via PACs (Political Action Committees) and think tanks whose sources of cash are kept confidential, you can't tell if there is bias or not.

people are attracted to," say the two now-wealthy ex-restaurant workers mentioned above.[141]

Shaq O'Neal's joking 2017 quip that "the earth is flat" can be taken seriously. There are now websites, blogs or social media sites advocating so many nonsensical or crazy opinions it's hard to imagine anything wilder. One site claims Australia doesn't exist, another that Finland is a fiction,[142] and another that Hillary Clinton is a lizard.[143]

How people, particularly North Americans, can fall for such "junk thought" is likely due in large part to a general "dumbing down" of the population, where "about a quarter of American adults (26 per cent) say they haven't read a book in whole or in part in the past year."[144] Author Susan Jacoby documents this cultural degradation thoroughly and devastatingly in her landmark book, *The Age of American Unreason*,[145] which describes the gradual rise, over roughly four decades, of "a new species of semiconscious anti-rationalism, feeding on and fed by an ignorant popular culture of video images and unremitting noise that leaves no room for contemplation or logic."

A more insidious and far more dangerous threat comes from "new software which lets users take existing videos and make high quality altered video and audio that appears real. The emergence of the technology opens up a new world of hoaxes driven by doctored audio or video, and threatens to shake faith in the media."[146] Warns U.S. Senator Ron Wyden: "I fully expect there will be a proliferation of these sorts of fictions to a degree that nearly drowns out actual facts."[147]

Whoever has the most money, tweets the most outrageously on Twitter, or hires the best fake video producers, can buy the most opinions. U.S. President Trump, and apparently Russian leader Vladimir Putin as well, know this. How far will they go to take advantage? No one can predict. Some social media propaganda campaigns are already government-sponsored.[148]

And Trump himself has openly teamed up with several notorious Science Denial Industry luminaries, openly bringing them into his government.

EPA wreckers

The actions of men like Myron Ebell and his fellows haven't been confined to defending tobacco, deriding endangered species rules or climate science. As a legislative assistant to Republican Representative John Shadegg, Ebell "advocated the elimination of federal conservation regulations, leaving protection of the environment in the hands of state and local officials."[149]

And in 2016, Ebell was chosen "to lead Donald Trump's U.S. Environmental Protection Agency (EPA) transition team."[150] The choice was typical of the new U.S. administration, which also named several other, similar players to what has been dubbed "the EPA wrecking crew." They included four men mentioned earlier in this chapter:[151]

- former tobacco industry attorney Steve Milloy, a one-time member with Ebell of the American Petroleum Institute's Global Science Communications Team, which "laid out the oil industry's strategy to undermine the science of global warming;"
- Craig Richardson, who has ties to Philip Morris, the Tobacco Institute and R.J. Reynolds Tobacco;
- Christopher Horner, a former attorney for R.J. Reynolds Tobacco and lobbyist for the Chemical Manufacturers' Association;
- James Enstrom, who has received funds from the Council for Tobacco Research, the Tobacco Institute, Philip Morris and R.J. Reynolds. He was widely criticized for failing to disclose his tobacco industry ties when he produced a study on second-hand smoke and mortality, and subsequently lost a job at UCLA.

Also part of the group was Amy Oliver Cooke, creator of a group called "Mothers in Love with Fracking."[152]

The entire Trump administration, in fact, soon became almost a Who's Who of climate change deniers or people with strong ties to the chemical or fossil fuel industries, starting with Secretary of State Rex Tillerson, former CEO of Exxon Mobil Corp., and Scott Pruitt, appointed head of the Environmental Protection Agency (EPA) despite having more common, and will increase in future, according to a variety of studies. One study at the University of Hawaii at Manoa found that a combination of high humidity and high heat were most lethal, with temperatures in the range of 30 C representing a general threshold where life is threatened. If the humidity climbs above 50 per cent, however, that threshold is reached at only 20 C.

Dr. Mora's group found that "currently, about 30 per cent of the world's population is exposed to the risk of lethal heat more than 20 days a year. Even if 'drastic reductions' of greenhouse gas emissions were applied, by 2100 almost half the world's population would face the equivalent risk."[168]

At the opposite extreme, there appears also to be a "remarkably strong correlation between a warm Arctic and cold winter weather further south," said Judah Cohen, a climatologist at Atmospheric and Environmental Research.[169] Where some heat waves may be induced by Jet Stream changes, the colder winters may be caused by disruption of the Polar Vortex, a low pressure system that moves around the pole region. Warmer polar temperatures could be weakening this movement, allowing the vortex to flow southward, dragging cold arctic air with it.

"The Arctic has just experienced its toastiest winter on record, with parts of the region 20C (68F) warmer than the long-term average, a situation scientists have variously described as 'crazy,' 'weird,' and 'simply shocking,'" reported *The Guardian*.[170] Meanwhile, "two large winter storms recently swept the U.S. east coast in less than a week, unloading up to three inches of snow per hour in places, resulting in several deaths, thousands of cancelled flights, closed schools and snarled traffic."

The events cited seem extreme to us today, but in coming decades may become routine. Our "new normal" extremes could reach proportions unimaginable now, with blistering heat, ferocious storms, sudden freeze-ups and waterfall-like downpours alternating with each other at dizzying speeds.

b) **Melting ice and snow-pack**, *as the cryosphere (frozen environment) gradually disappears:* Arctic sea ice and on-land glaciers will turn to liquid, with a variety of local and regional effects, including sea level rise and the flooding of coastal cities. The *Wikipedia* sums up the situation with on-land mountain glaciers briefly:

> *Since 1980, a significant global warming has led to glacier retreat becoming increasingly rapid and ubiquitous, so much so that some glaciers have disappeared altogether, and the existence of many of the remaining glaciers are threatened. In locations such as the Andes of South America and Himalayas in Asia, the demise of glaciers in these regions has the potential to affect water supplies in those areas ... The acceleration of the rate of retreat since 1995 of key outlet glaciers of the Greenland and West Antarctic ice sheets may foreshadow a rise in sea level, which could affect coastal regions.*[171]

Many lowland areas depend on annual glacial runoff to irrigate crops and keep local water reservoirs replenished. If glaciers disappear, they will be in trouble. Reduced glacial cold water runoff in spring can also lead to insufficient stream flows to allow freshwater fish species such as salmon and trout to survive and reproduce.

The global meltdown is obvious everywhere. In France, for example, "all six of the major glaciers in that country are in retreat. On Mont Blanc, the highest peak in the Alps, the Argentière Glacier has receded 1,150m (3,770 ft) since 1870. Other Mont Blanc glaciers have also been in retreat, including the Mer de Glace, which is the largest glacier in France at 12km (7.5 mi) in length but retreated 500m (91,600 ft) between 1994 and 2008 ... A Swiss glacier survey of 89 glaciers found 76 retreating, five stationary and eight advancing from where they had been in 1973."[172]

In the U.S., one news article warns "the iconic Glacier National Park will need to consider changing its name, as a recent report reveals that there are only 26 glaciers left there. In the latter 19th century, the park was home to 150 glaciers. The same report states that it **is 'inevitable' that the contiguous U.S. will lose all of its glaciers by 2050.**"[173]

impregnable deep-freeze to protect the world's most precious seeds from any global disaster and ensure humanity's food supply" was flooded in May 2017.[184] The Global Seed Vault, buried in a mountain inside the Arctic Circle, on the Norwegian island of Spitsbergen, was "breached after global warming produced extraordinary temperatures over the winter, sending meltwater gushing into the entrance tunnel." Luckily, the water didn't reach the seeds themselves.

c) **Widespread, longer-lasting droughts, and bigger wildfires**, *affecting increasingly larger regions:* Already-dry areas, such as the African Sahel, will expand, as desertification spreads; while abnormally severe or persistent dry spells will hit more temperate regions, giving rise not only to crop losses and soil erosion, but to larger, more frequent wildfire outbreaks. Grasslands, forests and cities are all vulnerable.

Technically, outdoor fires have different names depending on what they are burning: brush fires consume dead brush, grass fires ignite dry grass and forest fires burn trees. But all come under the general heading of wildfire. The alternating wet spell/dry spell character of global warming's pendulum swings can increase the risk, first by heavier-than-normal rainfall spurring the growth of thicker vegetation, then by drought drying that vegetation out, producing more fuel than usual for fires. As the *Wikipedia* notes: "Years of precipitation followed by warm periods can encourage more widespread fires and longer fire seasons."[185]

Any doubt that the wildfire phenomenon is increasing was dispelled in 2016-17, in both North America and Europe. The dramatic 2016 Fort McMurray Wildfire in Alberta, Canada, the disastrous 2017 fires in California, and across the Atlantic in 2017 in Portugal and in Siberia's boreal forests, all set records, as did the subsequent, unprecedented, mind-boggling 2019-2020 countrywide burning of Australia. So vast was the damage done Down Under that it may be years before an accurate assessment is complete, and this book consequently won't attempt it. Suffice to say that the entire country will be decades recovering from it.

The Fort McMurray fire raged from 1 May 2016 to 2 August 2017 (though considered "under control" by 5 July 2016, it "smoldered in deeper layers of moss and dirt throughout the winter"[186] and blazed up again the following year). Before it was fully extinguished, it had forced the evacuation of more than 90,000 people from their homes—the largest evacuation in the province's history—and destroyed 1.5 million acres, including 2,400 homes and buildings in the city of Fort McMurray itself. Causing insurance payouts of more than $4.7 billion, it was "the most expensive disaster in Canadian history."[187]

Wildfires also plagued neighbouring British Columbia in 2017, with 375 separate fires sweeping the province, "threatening timber supplies and sending lumber prices surging ... West Fraser Timber Co. suspended operations at three lumber mills that represent an annual production of 800 million board feet of lumber and 270 million square feet of plywood. Norbord Inc., the largest North American producer of oriented strand board used in residential construction, also suspended production at its mill in 100 Mile House in central B.C."[188] At least two mines also suspended production, while thousands of people were forced to evacuate as fire covered hundreds of square kilometres.

The smoke and particulates from the B.C. fires created "high-risk air quality levels"[189] in B.C. and neighbouring Alberta. Children, seniors, and those with cardiovascular or lung disease were advised to stay indoors.

The 2017 California fire season was "the most destructive wildfire season on record" with dozens of individual fires burning in both northern and southern parts of the state. Statewide, a total of 9,133 fires burned 1,381,405 acres, and killed at least 44 people [46 by later count]. Fires in December alone forced 230,000 people to evacuate.[190] Destroyed, along with thousands of other structures, was "Peanuts" cartoonist Charles Schultz's home near Santa Rosa.[191] Flames near the city of Ventura were clocked travelling at "an acre per second."[192]

A series of 156 fires that ripped through Portugal in June 2017 were even deadlier, killing at least 66 people and injuring 204. Most were killed "when a fire swept across a road filled with evacuees escaping in their cars."[193] Preceded by a heat wave whose temperatures reached 40C (104F), the fires caused "the largest loss of life due to wildfires in Portugal's history."[194]

Meanwhile, in the far north, wildfires have been plaguing the boreal forests of Siberia. A 2013 study published in the Proceedings of the National Academy of Sciences "determined that recent fires in the region's forests have been the worst in 10,000 years." They continued in later years, with at least 27,000 hectares (100 square miles) burning in 2017 in the Irkutsk region of southern Siberia and another 27,000 hectares burning in neighbouring states.[195]

Not far away, southern Russian, the Ukraine, Romania and Moldova were blanketed in "orange snow" in March 2018, as sands from sandstorms in the Sahara Desert blew far north and mixed with late-spring snow to create an "atmosphere from Mars."[196] Though blowing sand in the area has occurred before, meteorologists said this time "the concentrations of sand are much higher." So high that the dust caused a "light brown streak" in the clouds, visible from NASA satellites orbiting the earth.

Although global warming is expected overall to result in increased rainfall and flooding, drought-vulnerable regions will see the opposite, with already-arid areas expanding and borderline-dry ones becoming more arid.

For example, the 1998-2012 drought in the eastern Mediterranean area known as the "Levant," was "the worst drought of the past 900 years,"[197] according to tree-ring records recorded in the Old World Drought Atlas. Countries affected included Greece, Lebanon, Jordan, Syria and Turkey. "The Mediterranean is one of the areas that is unanimously projected [in climate models] as going to dry in the future," said Yochanan Kushnir of the Lamont Doherty Earth Observatory.[198]

The California drought that led to the state's record wildfires was "the worst drought in 1,200 years" in the region, according to the Natural Resources Defense Council (NRDC), which added "California's aquifers are overdrafted to the point of causing massive land subsidence that destabilizes local infrastructure,"[199] namely landslides.

The state had received an earlier taste of what was to come in 2009, when drought hit the once-fertile Central Valley, source of "more than a quarter of the nation's fruits, nuts and vegetables," reducing it to "a 21st century Dust Bowl."[200] The valley's bounty was based on its

70-year-old irrigation system, fed annually by meltwater from the Sierra Nevada range's melting snowpack. By the end of the century, scientists predicted then, the central part of the state could experience temperatures rivalling those of Death Valley, and face the loss of 90 per cent of its water source.

Meanwhile, in South Africa, Cape Town was preparing in the spring of 2018 to become the "first [major city] in the world to turn off its water taps," after its water supply was almost totally dried up by three consecutive years of "a drought so severe it would usually be expected only once every 384 years."[201] Authorities were gearing towards "Day Zero," when the six-dam reservoir system falls to 13.5 per cent of capacity. "In my 40 years in emergency services, this is the biggest crisis," said Greg Pillay, head of the city's disaster operations center.

According to a study published in *Nature Climate Change*, drought severity increased across the Mediterranean, southern Africa and the eastern coast of Australia during the 20th century, while semi-arid areas of Mexico, Brazil, southern Africa and Australia have started turning into desert as the world warms. "Our research predicts that aridification would emerge over about 20 to 30 per cent of the world's land surface by the time global mean temperature change reaches 2C," said study co-author Manoj Joshi of the University of East Anglia.[202]

That would be the grand daddy of all Dust Bowls, with orange snow for everyone ...

d) **Local flooding,** *as weather extremes, coupled with surplus water from melting glacial ice, cause sudden rises in lake and river levels:* Flash floods—as opposed to coastal flooding or general, region-wide water level rises—may become more common.

A graphic example was seen in the St. Lawrence seaway region in 2017, when record floods inundated Quebec. Record amounts of rain fell that year in Quebec and Ontario, saturating the soil till it could not absorb more. The runoff added "to water levels in already bloated rivers and streams," and "Lake Ontario hit its highest recorded level since record keeping began. The St. Lawrence is about 1.2 meters higher than it normally is this time of year."[203] Climate scientist Paul Beckwith of

falciparum and spread by the *Anopheles* mosquito, malaria has traditionally been confined to tropical or subtropical areas, but as currently temperate regions become warmer and more humid, will spread there. Projections are[216] that it could be present by 2050 in the U.S., including Florida, Texas and Washington State, and in areas of southern Brazil and eastern Europe where it is presently unknown. A similar development is foreseen for dengue fever, a virus transmitted by the *Aedes* mosquitos. Deadly Yellow fever is also mosquito-borne and will likely also spread northward, along with the equally nasty Rift Valley fever. Also spread by *Aedes* mosquitos is Zika virus, recently reported in Florida.[217]

Tick-borne diseases currently spreading to more temperate regions include sometimes fatal Lyme disease, now firmly established in southern Ontario, Canada, and the malaria-like disease babesiosis. Once limited to the tropics, the latter has recently cropped up "everywhere from Italy to Long Island, N.Y."[218] Also carried by ticks is the viral powassan disease, or POW, now spreading in the northeastern Great Lakes region of the U.S. and Canada.[219] Yet another threat is Chagas disease, spread not by ticks but by the so-called "kissing bug," which tends to bite sleeping humans around the mouth. Previously "unheard-of" in the U.S., the bugs are now common "across the bottom two-thirds of the U.S.," according to Patricia Dorn, of Loyola University.[220] Recent studies have found that half of the bugs carry *Trypanosoma cruzi*, the parasite that causes Chagas, and 38 per cent of bugs collected in Arizona and California also carried human blood.

Other dangers to humans that are expected to become more common in currently temperate regions include cholera, the deadly Ebola virus, plague, carried by fleas (yes, **plague**, as in the Black Death described in Chapter One), which may increase in the western U.S. and Canada,[221] sleeping sickness and both human and livestock varieties of tuberculosis. The flea that carries plague finds a home on the backs of rats, which themselves carry a variety of diseases, including human strains of the antibiotic-resistant staph bacteria MRSA, the diarrhoea-causing *Clostridium difficile*, and Leptospirosis, which is spread via rat urine. Also carried by rats, whose populations are booming in urban areas as the temperature warms, is Seoul virus, a type of hantavirus that can provoke kidney malfunctions.[222]

According to the Canadian Medical Association (CMA), rodent-carried pathogens "such as the Sin Nombre virus, a cause of hantavirus pulmonary syndrome, are harboured by rodent species that are especially abundant in the southwestern United States and in western states and provinces. Rodent population density appears to be a key driver of disease in humans, who are often infected after exposure to dust contaminated by rodent urine (e.g. while sweeping out grain storage spaces). Recent surges in hantavirus infection have been attributed to the El Nino-type weather conditions, which may become predominant with future climate change, suggesting that hantavirus pulmonary syndrome incidence may increase in coming decades.

Water-borne infectious agents, such as legionellosis (the cause of Legionnaires' disease) are also likely to increase. Legionnaires', for example, "peaks during warmer months and risk appears to increase with rainy, humid weather."[223] Invasive fungal diseases may spike, as well. "The dry summers and heavy wintertime precipitation projected for North America match the optimal conditions for spore dispersal by *Blastomyces dermatitidis*, a fungus that causes disease of the bones, lungs and skin. In Canada, *B. Dermatitidis* occurs most commonly in northwestern Ontario." And warmer, drier summers "may have facilitated the establishment of *Cryptococcus gattii* in Canada. This fungus had previously been seen only in tropical and subtropical regions (particularly those in Australia), but emerged on Vancouver island in 1999, where it has caused more than 100 cases of human illness in addition to illness in domestic animals."[224]

In addition to tuberculosis, livestock diseases expected to move to new regions include Avian Influenza (H5N1), commonly known as "bird flu," which can cause high mortality among domestic poultry. In early 2013 another strain (H7N9) was found to have spread to humans. "On 9 January 2017, the National Health and Family Planning Commission of China reported to WHO 106 cases ... including 35 deaths."[225]

Microbial species other than disease vectors will also be affected by climate change, including the soil bacteria that play a key role in soil nutrient content and agricultural productivity. Several recent studies have indicated that—despite rapid reproductive rates that would seem to allow them to respond quickly to climate shifts—soil bacteria will in fact be affected by climate change, but slowly, with time lags of three to 10 years or more. The authors of one study comment on this counter-intuitive time lag:[226]

> *By virtue of their short generation times and dispersal abilities, soil bacteria might be expected to respond to climate change quickly and to be effectively in equilibrium with current climatic conditions. However, legacy effects—defined here as community properties that persist after environmental change—have been observed in soil microbial communities, which take up to three years to respond to drought and other environmental shifts. There is an indication of decadal-scale legacy effects in microbial enzyme activity as well. Microbial legacy effects are also known in agricultural and other ecosystems. Moreover, because the distributions of soil bacteria are strongly influenced by edaphic characteristics (including soil pH and soil nutrient availability), and because these soil properties change slowly over time, factors driving shifts in soil bacterial communities can reflect historic climate.*
>
> *Thus, soil bacterial communities may still be adjusting to existing climate change, and it may take years or decades for the full effects of existing climate change to become evident.*

In short, the change may creep up on observers slowly, then come as a surprise, for which farmers may be unprepared.

Heat stress alone is expected to have a negative effect on livestock. "Thermal stress is one of the greatest climate challenges faced by domestic animals ... Feed consumption by dairy cattle starts to decline when average daily temperature reaches 25 to 27C and voluntary feed intake can be decreased by 10-35 per cent when ambient temperature reaches 35C and above. Mallonee et al reported that during hot weather feed consumption in cows is reduced by 56 per cent during the day time."[227] And, as every dairyman knows, less feed consumed means less milk produced.

Nor are birds, animals and humans the only targets. Staple farm crops like wheat, soybeans and corn are all threatened, as the *Wikipedia* notes:

> *While warmer temperatures create longer growing seasons, and faster growth rates for plants, it also increases the metabolic rate and number of breeding cycles of insect populations. Insects that previously had only two breeding cycles per year could gain an additional cycle if warm growing seasons extend, causing a population boom. Temperate places and higher latitudes are more likely to experience a dramatic change in insect populations. ... Studies found that soybeans with elevated CO_2 levels grew much faster and had higher yields, but attracted Japanese beetles at a significantly higher rate than the control field. The beetles in the field with increased CO_2 also laid more eggs on the soybean plants and had longer lifespans, indicating the possibility of a rapidly expanding population. De Lucia projected that if the project were to continue the field with elevated CO_2 levels would eventually show lower yields than that of the control field.*[228]

"Crops in the northern hemisphere will encounter pathogens they have not seen before," warned the *Scientific American*, reporting on a paper by fungal biologist Sarah Gurr, of the University of Exeter.[229] On average, "crop pests and pathogens in the northern hemisphere have moved northward by about 1.7 miles per year since 1960," said the article, particularly fungi or fungal-like organisms like oomycetes. It quoted study author Gurr: "If you are standing in a wheat field in North America or Canada, it's hectare upon hectare of genetically uniform wheat with perhaps one resistance gene to hold back the march of fungi. It will not necessarily have the right resistance genes for pathogens that are moving."

Meanwhile, tree-killing southern pine beetles are moving northward, attacking some of the most valuable timber species. "They're spreading farther north as global temperatures rise, putting entire ecosystems at risk and creating fuel for wildfires as they kill the trees they infest," reported *Inside Climate News*.[230] "A new study shows the insects' range could reach Nova Scotia by 2020 and cover more than 270,000 square miles of forest from the Upper Midwest to Maine and into Canada by 2080. Winter cold snaps that once killed the beetles in their larval stage are becoming less frequent at the northern edge of the beetles' current range, which will allow them to multiply and spread into new territory quickly."

And finally, wildlife may suffer, as "climate-driven increases in rabbit and predator populations (e.g. fox) may augment the risk of tularemia and rabies."[231] Migratory birds, on whom humans depend to keep many insect pest populations under control, are seeing their life cycles disrupted as local temperatures warm, then abruptly shift to unusual cold, then back to premature warming. Food sources that once supplied the birds on their travels, and nesting sites where they found forage, may be ready either before they arrive, or too late after arrival. "The growing mismatch means fewer birds are likely to survive, reproduce and return the following year," said Stephen Mayor, lead author of a study published in *Science Advances*.[232]

f) **Agricultural and timber growing season changes,** *favouring some crops and harming others.* Beech trees, for instance, are increasing throughout the northeastern U.S. and Canada, favored by warmer weather. A study of forest composition in New Hampshire, New York and Vermont, from 1983 to 2014, found beech abundance "increased substantially, but species that include the red maple, sugar maple and birch decreased."[233] Beech wood is of lower value as lumber than the species it replaces, but warmer temperatures and the fact that deer prefer to eat the seedlings of other trees, rather than beech, lead to its dominating stands. "For loggers and timberlands, more beech trees may mean less access to better lumber."

Not to mention less maple sugar and maple syrup as sugar maple tree numbers decline. The annual maple sugar harvest in Canada and the northern U.S., for example, may be adversely affected if the required seasonal alterations of heat and cold are seriously disturbed.

Orchard crops—apples, plums, pears, even warmer-area oranges, grapefruit—which flower and are pollinated in spring, while ripening in autumn, may find it difficult to adjust, leading to decreased yields. That is, if the destruction of pollinators described in Chapter Two hasn't already made their reproduction impossible.

Grain crops, corn and forage grasses could also face new difficulties, over and above new insect or fungal pests, as increasing temperatures and CO_2 levels "add a level of complexity to figuring out how to maintain sustainable agriculture."[234] For example, increasing CO_2 would logically be expected to increase crop yields, and to some extent this is true. But, as with Japanese beetles and soybeans, mentioned in the previous section, the overall results could be counter-intuitive. The Royal Society, in 2005, predicted that the benefits of rising CO_2 are "likely to be far lower than previously estimated when factors such as increasing ground-level ozone are taken into account."[235] Other studies show that increased CO_2 actually reduced the concentrations of some nutrients, such as zinc, iron and protein, in plants, especially in grasses and legumes, as well as rice. "Increases in CO_2 lead to decreased concentrations of micronutrients in crop plants. This may have a knock-on effect on other parts of ecosystems as herbivores will need to eat more food to gain the same amount of protein ... Reduced nitrogen content in grazing has also been shown to reduce animal productivity in sheep."[236] Poorer quality rice, in Asia, would obviously also affect that region's people.

Some crops, in short, may be more plentiful, but less nourishing.

And of course, if CO_2 rises boost crop growth, they will also boost the growth of weeds. "Since most weeds are C3 plants, they are likely to compete even more than now against C4 crops such as corn."[237]

Where natural systems are concerned, nothing is simple, and everything affects everything else.

g) **Human migration**, *brought on by drought, flooding or climate-induced hunger or poverty, could increase sharply, creating social, economic, political and military complications.* A report in *The Guardian*, quoting a World Bank study, speaks for itself:

commercial fishing is an industry that in 2007 was valued at $3.8 billion for the U.S. alone.[258]

For human lovers of seafood, acidification may even bring serious risks of poisoning. An increase in so-called "red tide" algal blooms is associated with acidity, and these blooms "contribute to the accumulation of toxins (demoic acid, brevetoxin, saxitoxin) in small animals such as anchovies and shellfish, in turn increasing occurrences of amnesic shellfish poisoning, neurotoxic shellfish poisoning and paralytic shellfish poisoning"[259] in humans who eat the anchovies or shellfish.

Nor are fish and other underwater dwellers the only creatures in danger. Some of Britain's seabird colonies, for instance, have undergone catastrophic losses due at least in part to global warming. "Our seabird colonies, especially those in northern Scotland, are withering away," said Euan Dunn, of the Royal society for the Protection of Birds (RSPB).[260]

On the rocky island of St. Kilda, for example, there has been "a 99 per cent reduction in kittiwake nests since 1990. Last year only one pair bred in all the monitored sites on the island, and that single chick died ... In addition, the kittiwake colony at Marwick Head on Orkney was once a noisy, bustling home to thousands of kittiwakes, but is today deserted ... Similarly, Fiar Isle puffins, with their large pale cheeks and brightly coloured bills, have dropped in numbers from 20,000 to 10,000 over the past 30 years, while on Orkney and Shetland guillemots have also halved in numbers with several colonies disappearing. Fulmars, arctic skuas and arctic terns have also suffered massive losses. Added together, these statistics indicate that, in the past 25 years, Scotland may have lost almost half its breeding seabird population." The losses may be chiefly due to lack of food, as zooplankton have declined dramatically, "while, further up the food chain sand eels—a critically important source for many birds—have disappeared from many parts of the Atlantic and North Sea."[261]

"All sea life will be affected by the increasing acidification," a recent study by the BIOACID Project concluded.[262] The speed at which the oceans are acidifying today is a key factor. The change over the past 200 years—roughly coinciding with human industrialization—has been faster than "any known change in ocean chemistry in the last 50 million years ... Such a relatively quick change doesn't give marine life, which evolved over millions of years in an ocean with a generally stable pH, much time to adapt."[263] Many, too many, won't adapt.

They may also find it difficult to adapt to the other half of the ocean/climate chemical equation, namely **deoxygenation**, also referred to as "ocean hypoxea" or "ocean suffocation."

Gases dissolve more slowly in hot solutions than in cool ones. This includes atmospheric oxygen, which dissolves in seawater more slowly as the water's temperature rises. Thanks to global warming, less oxygen moves from the air into ocean waters, with negative consequences for marine life. Making the problem worse is the fact that the metabolism of most creatures increases with heat, leading to faster respiration rates. That's why dogs (and some people) pant on really hot days.

A negative feedback loop sets in, putting the brakes on life. Fish, crabs and other marine organisms need more oxygen as the temperature rises, but because less is dissolving from the atmosphere, less is available. The law of supply and demand stops: creatures strain for more oxygen, but can't get it—they suffocate, and the number of creatures drops.

At the same time, warming causes the water to stratify, with a lighter, less salty layer buoying to the top, and prevents the upwelling of nutrients from deeper, heavier ocean layers. The top layer is where phytoplankton—microscopic plants, that take in CO_2 and give off oxygen via the process of photosynthesis—live. Lacking sufficient nutrients, their numbers thin, leading to less oxygen being produced to either mix back down into deeper layers where marine creatures can get it, or return to the atmosphere where surface animals, including people, need it to breathe. Yet another negative feedback loop.

The consequences of this are already being felt, as so-called "oxygen minimum zones," where life can't survive, spread throughout the world's waters.[264] One such zone has appeared off the coasts of Guatemala and Costa Rica, where researchers noticed that blue marlin, prized gamefish which normally dive as much as a half mile deep chasing prey, were found staying near the surface, "rarely dropping more than a few hundred feet." They were trying to avoid suffocation, researchers told

Eastern Seaboard could see an additional 30 inches of sea level rise as the backed-up currents pile water up on East Coast shores."[276]

Research into the potential effects of further ocean current disruption has barely begun, but the possibilities it has already revealed surely indicate that the Precautionary Principle should take over, fast.

j) **Methane release,** *from deposits currently trapped in the form of methane hydrate under arctic ice and permafrost:* One researcher calls this "the REALLY Big Kahuna." A rapid release of methane into the atmosphere 250 million years ago was likely the chief cause of the so-called Permian Mass Extinction Event, which destroyed "nearly 90 per cent of all the species on the planet."[277]

In today's atmosphere, methane (CH_4) is present as a mere "trace" gas, making up only 0.00019 per cent of dry air, by volume. But its small volume belies its potential influence, as well as its actual effects in past eras when it was present in far larger amounts. What makes methane important is its sheer strength as a heat-trapping greenhouse gas. Its heat-trapping potential is "84 times greater than CO_2 in a 20-year time frame,"[278] and may still trap 28 times more heat per mass unit than CO_2 for as long as 100 years after emission.

What's more, when it breaks down it produces CO_2 and water vapour—which both continue to trap heat. And the water vapour makes its way upward to higher atmospheric levels, where it forms both water and ice clouds in the stratosphere, preventing heat from rising further and enhancing the greenhouse effect below even more. Researchers fear large increases in methane may lead to "a warming that increases non-linearly"—in other words greater or faster than a direct-line relation with the amount would indicate.

There is evidence that the Permian Extinction Event was caused by "a runaway greenhouse effect"[279] resulting from the possible combination of a meteor impact, a massive volcanic eruption of the so-called Siberian Traps, and the release of huge amounts of methane by methane-producing microbes called methanogens. During the Event, "the oceans' surface temperatures reached 40C (104F). It was simply too hot for life to survive."[280]

Of course, most scientists wouldn't expect such a massive burst of methane to happen again due simply to present global warming mechanisms. The Permian-Triassic Event, after all, likely had the help of a meteor from outer space! But any relatively abrupt release of the methane currently held under Arctic permafrost would still do great damage, especially when combined with the overall effects of rising CO_2. And the concentration of methane in earth's atmosphere has already increased "by about 150 per cent since 1750," reaching its highest value "in at least 800,000 years."[281]

Some researchers believe the size of methane release global warming might cause is being underestimated, leading to complacency. Interestingly, they are those most familiar with Arctic conditions, and the forces in play in the region. They note that many estimates of potential methane release have focused chiefly on the gas contained under on-land permafrost, not taking into account that trapped under permafrost deposits under the northern seas. Many previous studies "completely ignore the new empirical evidence on permafrost-associated shallow water methane hydrates on the Arctic shelf."[282]

Swedish researchers who sailed Russia's northern coast in 2008, for example, found "areas of sea foaming with gas bubbling up through 'methane chimneys' rising from the sea floor. They believe the sub-sea layer of permafrost, which has acted like a lid to prevent the gas from escaping, has melted away to allow methane to rise from underground deposits formed before the last ice age."[283]

The seabed permafrost underlying the East Siberia Arctic Shelf (ESAS), for instance, was once thought impermeable. But Arctic specialists from the U.S. and Russia who surveyed the area in 20 field expeditions from 1999 to 2011 reported "extremely high concentrations of methane (up to 8 ppm) in the atmospheric layer above the sea surface," which they attributed to seabed permafrost layer "degradation and thawing."[284]

On land, in northwest Canada, permafrost decay has already affected 52,000 square miles, "an expanse the size of Alabama."[285] In Siberia, once-frozen ground is collapsing across wide areas of tundra, causing hills to crumble and forming "large new valleys and lakes."[286]

And the more methane is released, the warmer the atmosphere becomes, thawing more permafrost in a positive feedback loop that could feed on itself. When the effect of these land emissions is combined with that of emissions from undersea sources, a catastrophic release becomes much more plausible.

Concluded one report: "All this proves that the $60-trillion price-tag for Arctic warming estimated by the latest *Nature* commentary should be taken seriously ... rather than denounced on the basis of outdated, ostrich-like objections based on literature unacquainted with the ESAS."[287]

And finally, if this were not enough, the presence of methane also degrades the ozone layer, because when it breaks down into water vapour it cools the stratosphere, producing "a relative increase in ozone depletion in polar areas, and the frequency of ozone holes."[288] "Stratospheric cooling is a key factor in the development of ozone holes over the poles."[289] The ozone layer prevents "most harmful UVB wavelengths of ultraviolet (UV) light from passing through the earth's atmosphere. These wavelengths cause skin cancer, sunburn and cataracts ... as well as harming plants and animals."

An explosion of the methane time bomb would not be helpful ...

k) **Species extinction**, *including that of the human species:* Our planet's 6th Great Extinction, rivalling those of past ages, and "on a par with the catastrophic demise of the large dinosaurs 65 million years ago,"[290] is already well underway, from a variety of causes. Climate change is bound to accelerate the process, and in fact is doing so today. We are in the midst of what is called an "Extinction Event."

In the normal course of nature, a small number of species go extinct all the time, part of the so-called "background rate" of one to five species lost per year[291] that forms part of the process of "natural selection," first described by Darwin. Ordinarily, they are replaced or even outnumbered as new, better-adapted species come on-line. But not during an Extinction Event, generally defined as **"a widespread and rapid decrease in the biodiversity on earth,"**[292] in other words, a planet-wide, catastrophic loss. When one of these hits, it packs a wallop.

So far as we know from the study of the fossil record, there have been five previous such events over the ages, the best-known of which are probably the Permian-Triassic Event, 252 million years ago, which wiped out 90 to 96 per cent of all then-living species, and the Cretaceous-Palogene Event, 66 million years ago, which wiped out 75 per cent of species, including the dinosaurs. Such events usually occur when "a biosphere under long-term stress undergoes a short-term shock,"[293] the shock being on a truly large scale—for instance, a really big asteroid hits the planet, or sea level falls drastically, or widespread seismic activity leads to many, continued volcanic eruptions.

Our own event, the Holocene extinction[294] (also dubbed the Anthropocene, or man-made, extinction) has so far witnessed species loss "at over 1,000 times the background extinction rate since 1900"[295]—some believe closer "to 10,000 times the background rate"[296]—and most scientists agree that the long-term stress and short-term shock causing it are the same: human activity.

The reason for the disparity between loss estimates is that no one is really certain how many species of creatures—which would have to include insects and microbes—actually exist in today's world. About two million species have been documented so far, but this may barely scratch the surface. All scientists can do is look at what has been identified, note how many have been lost or are being lost, and extrapolate the percentages from there.

Numerous studies have documented the ongoing human-caused extinction of creature after creature, due to habitat destruction, pesticide overuse, land clearing for intensive farming, destruction of soil organisms by inorganic fertilizers, industrial-scale logging, and introduction of invasive species [including genetically manipulated organisms (GMOs)], which may displace or attack the native inhabitants of unprotected ecosystems.

Global warming makes many of these situations worse. Its overall effect is to reinforce the rush to destruction.

The first mammal species known to go officially extinct as a direct result of climate change was the Bramble Cay melomys, also called the mosaic-tailed rat, a tiny rodent that lived on an island off Australia.[297]

What's really happening?

In system terms, what's happening here is that the purpose of the energy system—which originally consisted of a campfire to keep us warm, cook our food and light the night, then grew through the stages of water and windmill power, through coal and steam to gasoline and diesel engines and then nuclear reactors, to power the increasingly sophisticated tools humans invented to do their work—has been perverted. Its purpose now is not to warm us, or do our work, but to make money, and then more money, and then more, for a constantly diminishing number of corporate investors and board members, many of whom are already millionaires or even billionaires. And to keep right on making that money—at any cost to other people, or to the other inhabitants of the earth.

This has brought the energy system, via an array of positive and negative feedback loops, into direct conflict with the planet's climate system. Worse, at least for us, it has also brought the corporate energy giants' wealthy directors into conflict with the human communication system—namely, language—by which we convey information to each other. The integrity of that system is fast being destroyed by the Science Denial Industry, and then by every half-baked crank who can sit at a computer and type out mental garbage on the Internet.

Put this together with the simultaneous destruction of the medical care system, the food production system and the human economic/political system, as described in Chapters One, Two and Three, as well as with several other possible disasters this book hasn't space to describe—such as nuclear war or nuclear power reactor meltdowns—and see what happens next.

That is, let's look at the cumulative force of **The Cascade Effect** of a collapse of everything, of all systems at once, each affecting all of the others and being affected by them, as well as by God alone knows how many other unforeseen disasters, multiplying each others' consequences 10-fold, 100-fold or 1,000-fold ...

This is what we currently face, along with the question, more urgent with every passing day, of what we can do about it.

1 Ann Piettrangelo, "Everything you need to know about lung cancer," *Healthline.com*, as posted online 16 April 2017, at www.healthline.com/health/lung-cancer
2 See *Wikipedia, the Free Encyclopedia* biographical entries under each name.
3 Sharon Lerner, "Republicans are using Big Tobacco's secret science playbook to gut health rules," 5 February 2017, *The Intercept*, as posted online 1 May 2017 at https://theintercept.com/2017/02/05/republicans-want-to-make-the-ep ..., 2-3.
4 See: George Monbiot, "The Denial Industry," 19 September 2006, *The Guardian*, as posted online 9 April 2017 at www.the guardian.com/environment/2006/sep/19/ethicalliving.g2, and *Rational Wiki*, "A comparative guide to science denial," as posted online 9 April 2017 at http://rationalwiki.org/wiki/A_comparative_guide_to_science_denial.
5 *Wikipedia, the Free Encyclopedia*, "Doubt is their product," as posted online 17 April 2017 at https://en.wikipedia.org/wiki/Doubt_Is_Their_Product, 1.
6 *Wikipedia, the Free Encyclopedia*, "Merchants of Doubt," as posted online 31 March 2017, at https://en.wikipedia.org,
7 Monbiot, *Op. Cit.*, 3.
8 Kenneth Kimmel, "Why scientists are fighting back. We've had enough of Trump's war on facts," 16 April 2017, *The Guardian*, as posted online 16 April 2017 at www.the guardian.com/commentisfree/2017/apr/15/why-scienti ..., 1-2.
9 *Loc. Cit.*
10 Sarah Kaplan, *The Washington Post*, "Soon it will be too late: scientists issue dire second notice to humanity," 13 November 2017, as reprinted in *The Toronto Star*, www.thestar.com/news/world/2017/11/13/soon-it-will-be-too-la ...
11 *Random House Webster's College Dictionary*, "Scientific method," (New York: Random House Inc., 1995) p. 1201.
12 Cristian Violatti, "Science," 28 May 2014, *Ancient History Encyclopedia*, as posted online 7 April 2017 at www.ancient.eu/science/, 2.
13 *Wikipedia, the Free Encyclopedia*, "History of science in early cultures," as posted online 7 April 2017 at https://en.wikipedia.org/wiki/History_of_science_in_early_cultures, 2.
14 *Ibid.*, 3.
15 *Wikipedia, the Free Encyclopedia*, "Scientific method," as posted online 2 April 2017 at https://en.wikipedia.org/wiki/Scientific_method, 5.
16 Andrew Zimmerman Jones, "Hypothesis, Model, Theory & Law," 2 October 2016, *thoughtco.com*, as posted online 18 April 2017 at www.thoughtco.com/scientific-hypothesis-theory-law-definitions-604138, 3.
17 Jones, *Op. Cit.*, 1.
18 *Wikipedia, the Free Encyclopedia*, "Null hypothesis," as posted online 24 May 2017 at https://en.wikipedia.org/wiki/Null_hypothesis, 6.
19 David Goodstein, "How science works," 11 February 2000, California Institute of Technology, as posted online 16 June 2017 at http://www.its.caltech.edu/%7Edg/HowScien.pdf, 3-4.
20 *Op. Cit.*, 16.
21 *Loc. Cit.*
22 *Wikipedia, the Free Encyclopedia*, "Primum non nocere," as posted online 16 June 2017 at https://en.wikipedia.org/wiki/Primum_non_nocere, 1.
23 *Wikipedia, the Free Encyclopedia*, "Precautionary principle," as posted online 3 April 2017 at https://en.wikipedia,.org/wiki/Precautionary_principle, 1.
24 Robert N. Proctor, "The history of the discovery of the cigarette-lung cancer link: evidentiary traditions, corporate denial, global toll," 1 March 2012, *BMJ Journals*, as posted online 16 April 2017, at http://tobaccocontrol.bmj.com/content/21/2/87.full, 1.

91 *Loc. Cit.*, Figure 11 "400,000 years of CO2 and temperature."
92 Tom Curtis, "Climate change cluendo: anthropogenic CO2," 25 July 2012, *Skeptical Science*, as posted online 11 August 2017 at https://skepticalscience.com/anthrocarbon-brief.html, 2.
93 *Skeptical science.com*, "How do human CO2 emissions compare to natural CO2 emissions?" as posted online 11 August 2017 at https://www.skepticalscience.com/human-co2-smaller-than-natural-emissions-intermediate.htm, 1.
94 Wolfson, *Op. Cit.*, 27.
95 *What's Your Impact*, "Main sources of carbon dioxide emissions," as posted online 7 August 2017 at https://whatsyourimpact.org/greenhouse-gases/carbon-dioxide-emissions, 5.
96 Catherine Brahic, "Climate myths: Human CO2 emissions are too tiny to matter," New Scientist, 16 May 2007, as posted online 7 August 2017 at www.newscientist.com/article/dn11638-climate-myths-human ..., 3.
97 Wolfson, *Op. Cit.*, 27.
98 Brahic, *Op. Cit.*, 3.
99 *Wikipedia, the Free Encyclopedia*, "Scientific opinion on climate change," as posted online 31 March 2017, at https://en.wikipedia.org/wiki/Scientific_opinion_on_climate_change, 1.
100 *Op. Cit.*
101 *Wikipedia, the Free Encyclopedia*, "Intergovernmental Panel on Climate Change," as posted online 19 August 2017 at https://en.wikipedia.org/wiki/Intergovernmental_Panel_on_Climate_ ..., 2.
102 Carbon Brief and Duncan Clark, "What is the IPCC?" 6 December 2011, *The Guardian*, as posted online 19 August 2017 at www.theguardian.com/environment/2011/dec/06/what-is-ipcc, 1.
103 *Op. Cit.*, 1.
104 *Loc. Cit.*
105 *Wikipedia, Op. Cit.*, "Intergovernmental Panel," 1.
106 Brief and Clark, *Loc. Cit.*
107 NASA, *Jet Propulsion Laboratory, Global Climate Change*, "Scientific consensus: Earth's climate is warming," 26 June 2017, as posted online at https://climate.nasa.gov/scientific-consensus/, 2 July 2017.
108 *The Governor's Office of Planning and Research*, "List of Worldwide Scientific Organizations," as posted online 24 August 2017 at www.opr.ca.gov/s_listoforganizations.php.
109 *Wikipedia, Op. Cit.*, "Scientific opinion on climate change," 15.
110 Shannon Hall, "Exxon knew about climate change almost 40 years ago," 26 October 2015, *Scientific American*, as posted online 28 March 2017 at www.scientificamerican.com/article/exxon-knew-about-climat ..., 3.
111 *Loc. Cit.*
112 Staff report, "Documenting electric utilities' early knowledge and ongoing deception on climate change," 26 July 2017, *Truth-out*, as posted online 29 July 2017 at www.truth-out.org/news/item/41408-utilities-had-early-knowle ..., 1-2.
113 John H. Cushman Jr., "Shell knew fossil fuels created climate change risks back in 1980s, internal documents show," 5 April 2018, *Inside Climate News*, as posted online 5 April 2018 at https://insideclimatenews.org/news/05042018/shell-knew-scientists-c ...
114 Benjamin Franta, "Shell and Exxon's secret 1980s climate change warnings," 19 September 2018, *The Guardian*, as posted online 19 September 2018 at www.theguardian.com/environment/climate-consensus-97-p ... ., 1-2.
115 *Loc. Cit.*
116 *Desmogblog*, "Myron Ebell," as posted online 31 August 2017 at www.desmogblog.com/myron-ebell, 1.

117 Myron Ebell, interview for the documentary "Everything's Cool: a Toxic Comedy About Global Warming," *City Lights Pictures*, 2007.
118 *Desmogblog*, Loc. Cit., "Myron Ebell."
119 *Wikipedia, the Free Encyclopedia*, "Myron Ebell," as posted online 9 April 2017, at https://en.wikipedia.org/wiki/Myron_Ebell, 3.
120 Hope Forpeace, "How Trump ally Myron Ebell spread misinformation for Big Tobacco and Big Oil," 28 September 2017, *Alternet*, as posted online 1 May 2017 at http://www.alternet.org/environment/how-trump-ally-myron-ebell-spread-misinformation-big-tobacco-and-big-oil, 1.
121 Frontiers of Freedom letter to Philip Morris, "Taxation, due process and the tobacco industry: a campaign proposal to stop a dangerous precedent," April 1998, Philip Morris Records, *University of California San Francisco Industry Documents Library*, as posted online 8 May 2017 at www.industrydocumentslibrary.ucsf.edu/tobacco/docs/#id=ptjj0046, 2.
122 *Op. Cit., Wikipedia*, "Myron Ebell," *3*.
123 *Wikipedia, the Free Encyclopedia*, "Tobacco Master Settlement Agreement," as posted online 17 April 2017 at https://en.wikipedia.org/wiki/Tobacco_Master_Settlement _Agreement, 1.
124 *Wikipedia, the free encyplopedia*, "Competitive enterprise Institute," as posted online 19 April 2017 at https://en.wikipedia.org/wiki/Competitive_Enterprise_Institute, 3.
125 Sharon Lerner, "Republicans are using big tobacco's secret science playbook to gut health rules," 5 February 2017, *The Intercept*, as posted online 1 May 2017 at https://theintercept.com/2017/02/05/republicans-want-to-make-the-ep ..., 2.
126 *Op. Cit., Wikipedia*, "Myron Ebell," 1.
127 *Wikipedia, the Free Encyclopedia*, "Information Council on the Environment," as posted online 28 August 2017 at https://en.wikipedia.org/wiki/Information_Council_on_the_Environment, 1.
128 *Loc. Cit.*
129 *Loc. Cit.*
130 Rafi Letzter, "Trump is taking advice on the future of the environment from a man who denies basic science," 21 November 2016, *Business Insider*, as posted online 5 January 2017 at www.businessinsider.com/trump-epa-climate-science-myron-ebe ..., 2.
131 Dana Nuccitelli, "New study uncovers the 'keystone domino' strategy of climate denial," 29 November 2017, *The Guardian*, as posted online 29 November 2017 at www.theguardian.com/environment/climate-consensus-97-per- ... 1.
132 Philip Bump, "Jim Inhofe's snowball has disproven climate change once and for all," 26 February 2015, *The Washington Post*, as posted online 8 September 2017 at www.washingtonpost.com/news/the-fix/wp/2015/02/26/jim ...
133 *All Fake Journals*, "All Fake Journals," 24 June 2017, as posted online at http://allfakejournals.blogspot.ca/2013/01/list-of-bogus-journals-fake- ...
134 *Phys.org*, "Predatory journals a global problem," 6 September 2017, as posted online at https://phys.org/news/2017-09-predatory-journals-global-problem.html
135 Marco Chown Oved, "Junk science publisher ordered to stop 'deceptive practices'," 23 November 2017, *The Toronto Star*, as posted online 24 November 2017 at www.thestar.com/news/world/2017/11/23/junk-science-publish ...
136 Suzanne Goldenberg, "Work of prominent climate change denier was funded by energy industry," 21 February 2015, *The Guardian*, as posted online at www.theguardian.com/environment/2015/feb/21/climate-chan ..., 1-2.
137 Upfront, "Exxon's funding of polar bear research questioned," 24 October 2007, *New Scientist*, as posted online 4 September 2017 at www.newscientist.com/article/mg19626273-300-exxons-fundi ..., 1.
138 Terrence McCoy, "How two unemployed guys got rich off Facebook, fake news

and an army of Trump supporters," 21 November 2016, *The Washington Post*, as posted online 21 November 2016 at www.thestar.com/news/world/2016/11/21/how-two-unemploy ..., 2-3.

139 *Ibid.*

140 Soroush Vosoughi, Deb Roy, Sinan Aral, "The spread of true and false news online," 9 March 2018, *Science*, Vol. 359, issue 6380, pp. 1146-1151.

141 McCoy, *Op. Cit.*, 2.

142 James Ball, "Australia doesn't exist! And other bizarre conspiracies that won't go away," 15 April 2018, *The Guardian*, as posted online 16 April 2018 at www.theguardian.com/technology/shortcuts/2018/apr/15/austra ...

143 JanFrel, "Inside the great reptilian conspiracy: from Queen Elizabeth to Barack Obama," 31 August 2010, *Alternet*, as posted online 19 April 2018 at www.alternet.org/story/147967/inside_the_great_reptilian_cons ...

144 Andrew Perrin, "Who doesn't read books in America?" 23 November 2016, *Pew Research Center*, as posted online 18 March 2018 at www.pewresearch.org/fact-tank/2016/11/23/who-doesn't-read-b ..., 1.

145 Susan Jacoby, *The Age of American Unreason*, New York: Pantheon Books, 2008, xi.

146 Ali Breland, "Lawmakers worry about rise of fake video technology," 10 February 2018, *The Hill*, as posted online 1 April 2018 at http://thehill.com/policy/technology/34320-lawmakers-worry-about-r ...

147 *Loc. Cit.*

148 Alex Hern, "Facebook and Twitter being used to manipulate public opinion," 19 June 2017, *The Guardian*, as posted online 19 June 2017 at www.theguardian.com/technology/2017/jun/19/social-media-p ...

149 Loc. Cit., *Wikipedia*, "Myron Ebell."

150 *Desmogblog, Op. Cit.*, "Myron Ebell," 5.

151 Lerner, *Op. Cit.* 2 - 5.

152 *Loc. Cit.*

153 Ledyard King, "Scott Pruitt on a mission to change the climate of the EPA," 28 November 2017, *USA Today*, as posted online 2 December 2017 at www.usatoday.com/story/news/politics/2017/11/26/scott-pruitt ...

154 Michelle R. Smith, "EPA keeps scientists from speaking about report on climate," 24 October 2017, *The Associated Press*, as posted online 24 October 2017 at www.thestar.com/news/world/2017/10/24/epa-keeps-scientists- ..., 1.

155 Oliver Milman, "EPA insiders bemoan low point in agency's history: 'People are so done."" 7 April 2018, *The Guardian*, as posted online 8 April 2018 at www.theguardian.com/us-news/2018/apr/07/epa-scott-pruitt-cri ..., 1.

156 Christine Todd Whitman, ""How not to run the EPA," 8 September 2017, *The New York Times*, as posted online at www.nytimes.com/2017/09/08/opinion/how-not-to-run-the-epa ..., 2.

157 Julian borger, "Trump will drop climate change from U.S. National Security Strategy," 18 December 2017, *The Guardian*, as posted online 18 December 2017 at www.theguardian.com/us-news/2017/dec/18/trump-drop-clima ...

158 Georgina Gustin, "Trump administration deserts science advisory boards across agencies," 18 January 2018, *Inside Climate News*, 2, as posted online 18 January 2018 at https://insideclimatenews.org/news/18012018/science-climate-change ...

159 Elizabeth Shogren, "Human role in climate change removed from federal science report," 6 April 2018, *Truth-out*, as posted online 6 April 2018 at www.truth-out.org/news/item/44080-wipeout-human-role-in-cli ..., 1.

160 Tess Riley, "Just 100 companies responsible for 71 per cent of global emissions, study says," 10 July 2017, *The Guardian*, as posted online at www.theguardian.com/sustainable-business/2017/jul/10/100-fo ...

161 Dana Nuccitelli, "March against madness–denial has pushed scientists out into

the streets," 25 April 2017, *The Guardian*, as posted online 25 April 2017 at www.theguardian.com/environment/climate-consensus-97-per- ..., 2.
162 Julie Dermansky, "Truitt's plan to debate climate science paused as science confirms human link to extreme weather," 22 December 2017, *Truthout*, as posted online 24 December 2017 at www.truth-out.org/news/item/43000-pruitt-s-plan-to-debate-cli ..., 2.
163 John Queally, "A storm that could 'rewrite history': Irma hits as most powerful Atlantic hurricane ever recorded," 6 September 2017, *Common Dreams*, as posted online 7 September 2017 at www.truth-out.org/news/item/41846-a-storm-that-could-rewrite- ..., 1.
164 Damian Carrington, "Mass die-off of sea creatures follows freezing UK weather," 5 March 2018, *The Guardian*, as posted online 17 March 2018 at www.theguardian.com/environment/2018/mar/05/mass-die-off- ..., 1.
165 Michael Slezak, "Third-hottest June puts 2017 on track to make hat-trick of hottest years," 19 July 2017, *The Guardian*, as posted online 20 July 2017 at www.,theguardian.com/environment/2017/jul/19/third-hottest-j ..., 2.
166 Jon Henley, "Extreme heat warnings issued in Europe as temperatures pass 40C," 4 August 2017, *The Guardian*, as posted online 4 August 2017 at www.theguardian.com/world/2017/aug/04/extreme-heat-warmi ..., 1.
167 Damian Carrington, "Climate change: 'human fingerprint' found on global extreme weather," 27 March 2017, *The Guardian*, as posted online 27 March 2017 at www.theguardian.com/environment/2017/mar/27/climate-chan ..., 1-2.
168 Ivan Semeniuk, "Risk of deadly heat waves to rise due to climate change: study," 19 June 2017, *The Globe & Mail*, as posted online 19 June 2017 at www.theglobeandmail.com/technology/science/earths-deadly-h ..., 1-2.r
169 Oliver Milman, "Extreme weather becoming more common as Arctic warms, study finds," 13 March 2018, *The Guardian*, as posted online 13 March 2018 at www. The guardian.com/environment/2018/mar/13/extreme-wint ..., 1-2.
170 *Loc. Cit.*
171 *Wikipedia the Free Encyclopedia*, "Retreat of glaciers since 1850," as posted online 20 March 2018 at https://en.wikipedia.org/wiki/Retreat_of_glaciers_since_1850, 1.
172 *Op. Cit.*, 2.
173 Dahr Jamail, "Scientists predict there will be no glaciers in the contiguous U.S. by 2050–but Trump is stomping on the gas pedal," 22 May 2017, *Truth-out*, as posted online 22 May 2017 at http://www.truth-out.org/news/item/40650-scientists-predict-there-wil ..., 2.
174 Lorraine Chow, "Alaskan glaciers have not melted this fast in at least four centuries," 13 April 2018, *Truth-out*, as posted online 13 April 2018 at http://buzzflash.com/commentary/alaskan-glaciers-have -not-melted-th ...
175 Philip Oltermann and Kate Connolly, "Melting glaciers in Swiss Alps could reveal hundreds of mummified corpses," 4 August 2017, *The Guardian*, as posted online 8 April 2017 at www.theguardian.com/world/2017/aug/04/melting-glaciers-swi ...
176 *Grist*, "On the rocks," #50, as posted online 8 November 2017 at http://grist.org/briefly/over-half-of-greenlands-ice-sheet-is-in-danger- ..., 1-2.
177 Dahr Jamail, "'Apocalyptic' melting transpires in Antarctica as earth wraps up a scorching year," 4 December 2017, *Truth-out*, as posted online 4 December 2017 at www.truth-out.org/news/item/42777-pollution-and-diseases-wre ..., 3.
178 Jonathan Watts, "Underwater melting of Antarctic ice far greater than thought, study finds," 7 April 2018, *The Guardian*, as posted online 2 April 2018 at www.theguardian.com/environment/2018/apr/02/underwater-me ..., 1.
179 *Loc. Cit.*
180 *Op. Cit., Wikipedia*, "Retreat of glaciers," 9.

change," 16 May 2017, *Truth-out*, as posted online 16 May a2017 at www.truth-out.org/news/item/40591-tick-borne-illnesses-could ...

220 Joshua E. Brown, "With climate change, U.S. could face risk from Chagas disease," 15 May 2012, *Science Daily*, as posted online 30 March 2018 at www.sciencedaily.com/releases/2012/03/120315140225.htm

221 Amy Greer, "Climate change and infectious diseases in North America: the road ahead," 11 March 2008, *Canadian Medical Association Journal*, as posted online at www.cmaj.ca/content/178/6/715.full

222 Wendy Leung, "Nature on the move," 7 July 2017, *The Globe and Mail*, as posted online 7 July 2017 at www.theglobeandmail.com/life/health-and-fitness/health/four-pest-and-plant-species-that-pose-health-risks-and-how-to-protectyourself/article35577281/, 2-3.

223 *Op. Cit.*, 719.

224 *Loc. Cit.*

225 *Wikipedia, the Free Encyclopedia*, "Avian influenza," as posted online 31 March 2018 at https://en.wikipedia.org/wiki/Avian_influenza, 1.

226 Joshua Ladau, Yu Shi, Xin Jing, Jin-Sheng He, Litong Chen, Xiangui Lin, Noah Fierer, Jack A. Gilbert, Katherine S. Pollard, Haiyan Chu, "Climate change will lead to pronounced shifts in the diversity of soil microbial communities," 1 September 2017, *bioRxiv Preprint*, as posted online 20 April 2018 at http://dx.doi.org/10.1101/180174

227 Yadav Sharma Bajagai, "Impacts of climate change on dairy cattle," October 2012, *Nepalese Veterinary Journal*, vol. 30, pp. 2-16, as posted online 30 March 2018 at www.foodandenvironment.com/2012/10/impacts-of-climate-ch ...

228 *Wikipedia, the Free Encyclopedia*, "Climate change and agriculture," as posted online 30 March 2018 at https://en,wikipedia.org/wiki/Climate_change_and_agriculture, 2.

229 Stephanie Paige Ogburn, "Crop pests on the move due to climate change," 3 September 2013, *Scientific American*, as posted online 30 March 2018 at www.scientificamerican.com/article/crop-pests-on-the-move-d ..., 3.

230 Bob Berwyn, "tree-killing beetles spread into northern U.S. forests as temperatures rise," 28 August 2017, *Inside Climate News*, as posted online 9 February 2017 at https://insideclimatenews.org/news/28082017/southern-pine-beetles-s ...

231 Greer, *Op. Cit., CMAJ,* 720.

232 Don Zukowski, "Climate destabilization is causing thousands of new species migrations: plant, animal, insect, bird," 26 June 2017, *Truth-out*, as posted online 26 June 2017 at www.truth-out.org/news/item/41064-climate-destabilization-cau ..., 1.

233 Alan Adamson, "Beech trees are booming in forests because of climate change," 26 February 2018, *Tech Times*, as posted online 31 March 2018 at www.techtimes.com/articles/221889/20180226/beech-trees-are ...

234 Op. Cit., *Wikipedia*, "Climate change and agriculture," 2.

235 *Op. Cit. Wikipedia*, "Climate change and agriculture," 6.

236 *Loc. Cit.*

237 *Loc. Cit.*

238 Fiona Harvey, "Climate change soon to cause mass movement, World Bank warns," 19 March 2018, *The Guardian*, as posted online 19 March 2018 at www.theguardian.com/environment/2018/mar/19/climate-chan ..., 1.

239 Craig Welch, "Climate change helped spark Syrian war, study says," 2 March2015, *National Geographic*, as posted online March 2018 at https://news.nationalgeographic.com/news/2015/03/150302-syria-war-climate-change-drought/

240 *Op. Cit.*, 2.

241 Ruth Pollard, "The Mediterranean refugee crisis explained," 26 August 2015, *Sydney Morning Herald*, as posted online April 2018 at www.smh.com.au/world/the-mediterranean-refugee-crisis-explained-20150826-gj7pgz.htm,1-2.
242 Damian Carrington, "Climate change will stir 'unimaginable' refugee crisis, says military," 1 December 2016, *The Guardian*, as posted online 7 April 2018 at www.theguardian.com/environment/2016/dec/01/climate-chang ...
243 *Wikipedia, the Free Encyclopedia*, "Environmental migrant," as posted online 7 April 2018 at https://en.wikipedia.org/wiki/Environmental_migrant
244 *Loc. Cit.*
245 *Loc. Cit.*
246 *Loc. Cit.*
247 *Wikipedia, the Free Encyclopedia*, "Ocean acidification," as posted online 8 April 2018 at https://en.wikipedia.org/wiki/Ocean_acidification, 1.
248 The Ocean Portal Team, "Ocean acidification," 2017, *Smithsonian Ocean Portal*, as posted online 8 April 2018 at http://ocean.si.edu/ocean-acidification, 5.
249 *Loc. Cit.*
250 Dahr Jamail, "Eight-year study confirms 'all sea life will be affected by acidic oceans," 25 October 2017, *Truthout*, as posted online 25 October 2017 at www.truth-out.org/news/item/42361-eight-year-study-confirms-a ..., 1.
251 Ben Smee, "Great Barrier Reff: 30 per cent of coral died in 'catastrophic' 2016 heatwave," 18 April 2018, *The Guardian*, as posted online 18 April 2018 at www.theguardian.com/environment/2018/apr/19/great-barrier-r ..., 2.
252 *Ibid.*
253 *Loc. Cit.*
254 Op. Cit., *Wikipedia*, "Ocean acidification," 4.
255 Op. Cit., *Ocean Portal Team*, 6-7.
256 *Ibid.*
257 *Loc. Cit.*
258 Op. Cit., *Wikipedia*, "Ocean acidification," 4.
259 Op. Cit., *Wikipedia*, Ocean acidification," 4.
260 Robin McKie, "Britain's seabird colonies face catastrophe as warming waters disrupt their food supply," 20 August 2017, *The Guardian*, as posted online 20 August 2017 at www.theguardian.com/environment/2017/aug/20/seabird-colon ..., 1-2.
261 *Loc. Cit.*
262 *Op. Cit.*, Dahr Jamail, "Eight-year study."
263 *Op. Cit.*, Ocean Portal Team, 1.
264 *Wikipedia, the Free Encyclopedia*, "Ocean deoxygenation," as posted online 14 April 2018 at https://en.wikipedia.org/wiki/Ocean_deoxygenation
265 Craig Welch, "Climate change is suffocating large parts of the ocean," 4 January 2018, *National Geographic*, as posted online 4 January 2018 at https://news.nationalgeographic.com/2018/01/climate-change-suffocating-low-oxygen-zones-ocean/, 1-2.
266 *Loc. Cit.*
267 *Ibid.*
268 Natalya Gallo, ""Ocean deoxygenation," 2018, *Ocean Scientists*, as posted online 15 April 2018 at http://oceanscientists.org/index.php/topics/ocean-deoxygenation, 2.
269 *Wikipedia, the Free Encyclopedia*, "Gulf Stream," as posted online 17 April 2018 at https://en.wikipedia.org/wiki/Gulf_Stream, 1-2.
270 *Wikipedia, the Free Encyclopedia*, "Humboldt Current," as posted online 17 April 2018 at https://en.wikipedia.org/wiki/Humboldt_Current, 1.
271 *National Oceanic and Atmospheric Administration*, "Ocean currents," as posted online 15 April 2018 at www.noaa.gov/resource-collections/ocean-currents, 1.
272 Erik Betz, "Climate change is weakening a crucial ocean current," 11 April 2018,

Discover Magazine D-brief, as posted online 15 April 2018 at http://blogs.discovermagazine.com/d-brief/2018/04/11/0cean-current- ...

273 *Loc. Cit.*

274 Damian Carrington, "Avoid Gulf Stream disruption at all costs, scientists warn," 13 April 2018, *The Guardian*, as posted online 13 April 2018 at www.theguardian.com/environment/2018/apr/13/avoid-at-all-c ...

275 *Op.Cit.*, Betz, *D-brief.*

276 Nicola Jones, "How climate change could jam the world's ocean circulation," 6 September 2016, *Yale Environment 360*, Yale School of Forestry & Environmental Studies, as posted online 15 April 2018 at https://e360.yale.edu/features/will_climate_change_jam_the_global_ ..., 3.

277 Dahr Jamail, "Release of arctic methane 'may be apocalyptic,' study warns," 23 March 2017, *Truthout*, as posted online 3 March 2017 at www.truth-out.org/news/item/39957-release-of-arctic-methane ..., 1.

278 *Wikipedia, the Free Encyclopedia*, "Atmospheric methane," as posted online 7 April 2018 at https://en.wikipedia.org/wiki/Atmospheric_methane, 7.

279 *Wikipedia, the Free Encyclopedia*, "Permian-Triassic extinction event," as posted online19 April 2018 at https://en.wikipedia.org/wiki/Permian-Triassic_extinction_event, 1.

280 *Op. Cit., Wikipedia*, "Permian-Triassic," 6.

281 *Op. Cit. Wikipedia*, "Atmospheric methane," 1-2.

282 Nafeez Ahmed, "Seven facts you need to know about the Arctic methane timebomb," 5 August 2013, *The Guardian*, as posted online 7 April 2018 at www.theguardian.com/environment/earth-insight/2013/aug/05 ..., 2.

283 Steve Connor, "Exclusive: the methane time bomb," 22 September 2008, *The Independent*, as posted online 7 April 2018 at www.independent.co.uk/environment/climate-change/exclusive ..., 1.

284 *Op. Cit.*, Ahmed, *Guardian,* 1-2.

285 Bob Berwyn, "Massive permafrost thaw documented in Canada, portends huge carbon release," 28 February 2-17, *Insideclimate News*, as posted online 1 March 2017 at https://insideclimatenews.org/news/27022017/global-warming-permaf ..., 2.

286 *Op. Cit.*, 4.

287 *Op. Cit.*, Ahmed, *Guardian*, 4

288 *Wikipedia, the Free Encyclopedia*, "Ozone depletion," as posted online 19 April 2018 at https://en.wikipedia.org/wiki/Qzone_depletion, 9.

289 Government of Canada, "Ozone depletion and climate change," as posted online 19 April 2018 at www.canada.ca/en/environment-climate-change/services/air-po ..., 2.

290 Sarah Zielinski, "Climate change will accelerate earth's sixth mass extinction," 30 April 2015, *Smithsonian Magazine*, as posted online 7 April 2018 at www.smithsonianmag.com/science-nature/climate-change-will ..., 1.

291 Noah Greenwald, "The extinction crisis," *Biological Diversity*, as posted online 21 April 2018 at www.biologicaldiversity.org/programs/biodiversity/elements_of ..., 1.

292 *Wikipedia, the Free Encyclopedia*, "Extinction event," as posted online 21 April 2018 at https://en.wikipedia.org/wiki/Extinction_event, 1.

293 *Op. Cit.*, 4.

294 *Op. Cit., Wikipedia*, "Extinction event," 2.

295 *Loc. Cit.*

296 *Op. Cit.*, Greenwald, 1.

297 Brian Clark Howard, "First mammal species goes extinct due to climate change," 14 June 2016, *National Geographic*, as posted online 12 April 2018 at https://news.nationalgeographic.com/2016/06/first-mammal-extinct-climate-change-bramble-cay-melomys/

298 The Long Read, "A different dimension of loss: inside the great insect die-off,"

14 December 2017, *The Guardian*, as posted online 5 March 2018 at www.theguardian.com/environment/2017/dec/14/a-different-d ..., 3

299 615 Patrick Barkham, "Europe faces 'biodiversity oblivion" after collapse in French birds, experts warn," 21 March 2018, *The Guardian*, as posted online 21 March 2018 at www.theguardian.com/environment/2018/mar/21/europe-faces ..., 1.

300 *Op. Cit.*, 2.

301 *Loc. Cit.*

302 Justin McCurry, "Red list: thousands of species at risk of extinction due to human activity," 5 December 2017, *The Guardian*, as posted online 5 December 2017 at www.theguardian.com/environment/2017/dec/05/red-list-thou ..., 1-2.

303 *Loc. Cit.*

304 *Op. Cit.*, Long Read, 5-6.

305 *Wikipedia, the Free Encyclopedia*, "Extinction risk from global warming," as posted online 21 April 2018 at https://en.wikipedia.org/wiki/Extinction_risk_from_global_warming, 1.

306 Robinson Meyer, "Human extinction isn't that unlikely," 29 April 2016, *The Atlantic*, as posted online 7 April 2018 at www.theatlantic.com/technology/archive/2016/04/a-human-ext ...,1-2.

Part II
WHEN IT COMES TOGETHER

Chapter Five

THE CASCADE EFFECT

"One day, I wrote her name upon the strand,
But came the waves, and washed it away ..."
—Edmund Spenser, *Amoretti*

Looking only at the effects of climate change, much of what our species has written upon "the sands of time" could be washed away, or burnt to ashes, or turned to dust in drought and blown off on the wind—dust, ashes and sand together—starting within these next few decades. That's bad enough, but then factor in all of the other system breakdowns:

- a medical system that now deliberately addicts and kills people for money, intentionally frustrates the search for cures to disease, and prices already-existing cures out of reach of the sick;
- a food production system that destroys the nutritional value of what it produces, while eliminating the pollinators and soil organisms that make all food production possible, and at the same time destroying the genetic sources, both wild and domestic, of crop varieties we may need to survive;
- an economic/political system that turns society's young people into permanent debt slaves, unable to lift their heads out of the student loan sharks' mire, waiting with their parents for a financial Greatest of All Depressions (GAD) engineered by banks, lenders and their political enablers;
- plus whatever other systems go down, from nuclear power meltdowns or nuclear war, to a "defense" or police system turning its guns against its own people, or against the coming waves of

What is to prevent the "extended family" of armed guards from popping a bullet into each billionaire head and taking the burrows over for themselves isn't mentioned. One stock brokerage house CEO told a journalist he did in fact wonder: "How do I maintain authority over my security force after the Event?"[5] The Event being a "euphemism for the environmental collapse, social unrest, nuclear explosion, unstoppable virus or Mr. Robot hack that takes everything down." He and his ilk also pondered "how would they pay the guards once money was worthless? What would stop the guards from choosing their own leader? The billionaires considered using special combination locks on the food supply that only they knew. Or making guards wear disciplinary collars of some kind in return for their survival. Or maybe building robots to serve as guards."

Also unmentioned in the missile site ad was the next step after the five-year supply of freeze dried food runs out, or the possibility that the sheer grotesqueness of living in a hole with simulated outdoor views, working out with weights and swimming in an indoor pool, while most of life on earth is dying just above your head might drive you even battier than you already are, to even contemplate such a thing. Especially if you reflect on the fact that you and your corporate board's policies knowingly caused it all. The ads don't mention any luxury, gold-framed mirrors in the burrows, where billionaires can look at their own faces and, perhaps, puke. Or smile, and say "FUDS!"

Black comedy indeed.

Not comic

But there's not much comedy in the likely process of what these clowns believe they will escape. Since it's already begun, it will at first involve only an increase in tempo and magnitude. How long will it continue to be gradual, background-level, day-to-day, until at some point what's going on becomes downright intrusive, obvious, noticeable to everyone?

More sensitive organisms will pick it up earlier, bit-by-bit, and then the less sensitive will notice, then the others, even the preoccupied or stupid. Like smoke in a darkened theatre from a single cigarette butt, spreading as it ignites first an adjacent candy wrapper, or single seat cushion, smoldering, then sending up a small flame, then a big one, till someone yells "FIRE!" What will the trigger be? No one really knows. But everything will be affected, and everything will affect everything else, spark after spark.

We've already had Fort McMurray and summer in California. When will we notice the really BIG hurricane, the floods in Bangladesh spreading to the shores of the Mississippi, or the Yangtze, the crumbling mountains, landslides, as much dust in our eyes as the Joad family felt in Steinbeck's classic 1939 tale of the Dirty Thirties?[6]

How will impoverished populations react, with no jobs or resources, no resistance to the new diseases, bacterial, viral, fungal, cropping up on all sides, against which medicine has nothing to offer, no new concoction that isn't priced beyond reach, or addictive and leading to oblivion. Weakened by self-inflicted food shortages, politically divided, without a plan, without a way out, the Pharisaical "elect," loud prayers notwithstanding, all flushing down the same drain with the rest of us publicans and sinners ...

Did John have a prescient glimpse of the 21st and 22nd centuries, sitting on Patmos Island so many years ago? An enlightenment experience? Were the images of his Apocalypse meant only to describe the Roman Empire of his day, and the interior spiritual lives of his contemporary readers, or also the empires and readers of a much, much later date? Is it presumption to think so? Yeshua clearly warned: "Ye know not the day, nor the hour."

But predicted by Scripture or not, do we know enough, or care enough, to at least try and spare our own children?

Start with what we know now, and extrapolate from there, recalling those first three axioms behind Systems Theory: 1) Everything is connected to everything else; 2) You can never do just one thing, and 3) There is no "away."[7]

Of these latter, *resistant strains of Staphylococci and Salmonella bacteria have already been found.*[12]

Recently, *Mycoplasma bovis*, a bacteria that causes mastitis, pneumonia, arthritis and other diseases in cattle, was found for the first time in New Zealand.[13] It was previously confined to Europe and the U.S. More than 150,000 cows will have to be slaughtered to prevent the spread of the infection. If resistant strains of common bacteria like this start to show up in cattle around the world, such drastic culls may become commonplace.

As for poultry and pet birds, they are prone to numerous bacterial infections, including gram-negative *Klebsiella, Pseudomonas, Aeromonas, entyerobacter, Proteus, Citrobacter, Escheria coli* and *Serratia marcescens* species, and gram-positive *Staphylococcus aureus, S. Intermedius, Clostridium, Enterococcus* and *Streptococcus* species. Of these, methicillin-resistant *S. Aureus (MRSA)*—infamous as a problem for humans—has already been found in birds. Several *Mycobacterium* infections found in birds have also developed resistance to antibiotic treatment.

In fish, infections of *Aliivibrio salmonicida* have been found to be resistant to drug therapy. Some *Mycobacterium* species that attack fish can also infect humans.[14]

How long before resistant varieties of all or most of these appear? No one knows. But the truly frightening fact is, a number of genetic means exist by which one resistant bacterial species can transmit its resistance to other species.

The enzyme NDM-1, mentioned on the *Wikipedia* list previously cited, is one such agent. Originally from India, it was first detected in 2008 in the bacteria *Klebsiella pneumoniae*, and subsequently spread to the UK, U.S., Canada and Japan. It can transfer between bacterial species, making them resistant to a range of beta-lactam antibiotics. Worse is a resistant gene found in the U.S. in 2016, MCR-1, which makes the common bacterium *E. Coli* resistant to many antibiotics, including the so-called "last line of antibiotic defense," colistin. If this spreads to other species, it could make antibiotics all but worthless.

Antibiotic resistant infections pose "a fundamental threat to society," warned former UN Secretary General Ban Ki-moon.[15] Resistant "superbugs" already cause "an estimated 700,000 deaths annually, and that number could top 10 million by 2050."[16]

England's chief medical officer, Dame Sally Davies, warns that, if antibiotics continue to lose their effectiveness, it would be "the end of modern medicine."[17] With no drugs to fight infection, common medical interventions such as caesarean sections, certain cancer treatments and hip replacement surgery will become incredibly risky, and transplant surgery would be "a thing of the past. We are really facing—if we don't take action now—a dreadful post-antibiotic apocalypse. I don't want to say to my children that I didn't do my best to protect them and their children."

A *Guardian* report adds:

> *An example is provided by transplant surgery. During operations, patients' immune systems have to be suppressed to stop them rejecting a new organ, leaving them prey to infections. So doctors use immuno-suppressant cancer drugs. In future, however, these may no longer be effective.*
>
> *Or take the example of more standard operations, such as abdominal surgery or the removal of a patient's appendix. Without antibiotics to protect them during those procedures, people will die of peritonitis or other infections. The world will face the same risks as it did before Alexander Fleming discovered penicillin in 1928. Routine surgery, joint replacements, caesarean sections, and chemotherapy also depend on antibiotics and will also be at risk ... Common infections could kill again.*[18]

Humans and animals would both fall victim.

As noted, all of these microscopic, disease-causing creatures are developing or have developed resistance to antibiotic treatment, in significant part due, as described in Chapter Two, to the misuse/overuse of antibiotics in the factory farm system (including fish farms), as well as to over-prescription in human medicine. And also, as explained in Chapter One, because the costs of developing new antibiotics is considered too high by the major pharmaceutical corporations.

"There has not been a new class of antibiotics introduced in over 30 years."[19]

"The economic models of for-profit pharmaceutical companies have not favoured the development of new treatments for infectious diseases," writes Leah Cowen, chair of molecular genetics at the University of Toronto Faculty of Medicine.[20] "There are many reasons pharmaceutical companies aren't doing this work. Treatment times are typically short compared to drugs for chronic diseases like diabetes or heart disease and microbial resistance can quickly render new drugs obsolete."

Not only are no new antibiotics being developed, but those that already exist are increasingly hard to come by, due to supply chain problems. For example, the common antibiotic benzathine penicillin G—the only drug that can prevent and treat the transmission of syphilis from mother to child—"was found to be unavailable in 39 countries in 2015, including India, Australia and the U.S."[21] According to a report by the Dutch group Access to Medicine, "antibiotic supply is patchy, complex and at risk of collapsing," in great part because "low profits mean low production levels."

Of course, some bacterial threats to humans and animals might be amenable to treatment by bacteriophages, those beneficial viruses described in Chapter One, on which no one holds any patent—and which are thus of no interest to the major pharmaceutical corporations, who see only modest profits, if any, in their sale. Will the resistance of these corporations to phage re-introduction in North America and other parts of the world, where they were once in common use, continue? Will the fact that either developing new antibiotics or re-introducing phages, though returning relatively little profit, might save thousands, perhaps millions, of human and animal lives be seen as important by the directors of those firms?

As the saying goes, three guesses, and the first two don't count.

An indication of the general moral values governing Big Pharma and its regulators, especially in the U.S., was noted by *Pro Publica's* Caroline Chen, who reports[22] that the U.S. Food and Drug Administration (FDA) is "increasingly green-lighting expensive drugs despite dangerous or little-known side effects and inconclusive evidence that they curb or cure disease. Once widely assailed for moving slowly, today the FDA reviews and approves drugs faster than any other regulatory agency in the world ...

"As patients (or their insurers) shell out tens or hundreds of thousands of dollars for unproven drugs, manufacturers reap a windfall. For them, expedited approval can mean not only sped-up sales, but also—if the drug is intended to treat a rare disease or serve a neglected population—FDA incentives worth hundreds of millions of dollars."

Chen cited several examples, including the drug Nuplazid, noting that, since it "went on the market in 2016 at a price of $24,000 a year, there have been 6,800 reports of adverse events for patients on the drug, including 887 deaths as of this past March 31."

New ranges

But what is equally important, many bacteria—both resistant and non-resistant species—are now **increasing their geographic ranges due to human-induced climate change**, coming into contact with humans, animals or even some crops unused to their presence, and hence not naturally immune to them.

Thanks to prominent media attention, the best-known bacteria whose range is widening are *Borrelia burgdorferi* and *B. Mayonii*, which cause Lyme disease. Carried and transmitted by ticks, their North American habitat was once limited to the U.S., particularly southern New England, the South Atlantic states and Wisconsin. However, "[o]wing to climate change, the range of ticks able to carry Lyme disease has expanded from a limited area of Ontario to include areas of southern Quebec, Manitoba, northern Ontario, southern New Brunswick, southwest Nova Scotia and limited parts of Saskatchewan and Alberta, as well as British Columbia. Cases have been reported as far east as the island of Newfoundland."[23] It is expected that the ticks range "will expand into Canada by 46 km per year over the next decade, with warming climatic temperatures as the main driver of increased speed of spread."

Also carried by ticks, and also expanding their ranges, are the potentially fatal Rocky Mountain Spotted Fever, caused by the bacteria *Rickettsia rickettsii*, and Ehrlichiosis, caused by a number of bacterial species in the *Anaplasmataceae* family. Both of these are turning up earlier and more often in the U.S. Midwest, according to Mary Anne Jackson, of Children's Mercy Hospital in Kansas City, MO. Ehrlichiosis used to be called "summer flu," she noted, but "now it can occur any time of the year."[24] Neither of these is yet known to be resistant to antibiotics, but if they become resistant could pose serious problems. So could that ancient scourge, cholera, caused by the bacterium *Vibrio cholerae*, which has also been expanding its range.

Crops are less susceptible to bacteria than people or animals, but some bacterial diseases of plants, prevalent today chiefly in subtropical or tropical regions, may start turning up further north. Crown gall disease, caused by *Agrobacterium*, or disorders caused by *Burkholderia*, *Proteobacteria*, or *Xanthomonas* species could extend their ranges. Will some also become resistant to known chemical treatments? Rice, soybeans, cassava, beans, cotton, tomato, pepper, carrot and other crops are all susceptible to Bacterial Leaf Blight, while cucumbers, beans, cereals, pepper and tomato are attacked by Bacterial Leaf Spot.

~

And the danger from bacterial infections is only part of the story, where human, animal and crop health is concerned. There are dozens of disease-causing viral, fungal and parasitical species, hitherto limited to tropical or semi-tropical regions—but no longer—waiting to spread over the earth. In fact, there may be scores, if evolution and changing environmental conditions lead to mutations. Dealing with the 10 plagues of Egypt could look like child's play in comparison.

Some recent examples of viral diseases whose range has expanded would include mosquito-borne West Nile Fever and the Zika virus, both of which caused excitement when they appeared moving north through the U.S. and Canada. Still waiting for their northern debuts are the viral Dengue Fever and once-tropical Yellow Fever. Hantavirus, carried by mice and rodents, appeared in the U.S. in 1993 "after a six-year drought ended with heavy snows and rainfall."[25]

Malaria, caused by the *Plasmodium* parasite, is spreading, as is Valley Fever, caused by the *Coccidioides* fungus, and the "brain-eating amoeba" *Naegleria fowleri*. Minnesota had its first ever infection of the latter in 2010.

Viral and other diseases are also, like their bacterial counterparts, able to develop resistance to current standard treatments. "Antivirals are used to treat HIV, hepatitis B and C, influenza, herpes viruses, including *varicella zoster* virus, cytomrgalovirus and Epstein-Barr virus. With each virus, some strains have become resistant to the administered drugs."[26] Infection-causing fungi are also developing resistance. These include *Cryptococcus neoformans* and *Aspergillus fumigatus*. The malaria parasite, mentioned above, has also developed resistance, as have the parasitic protozoa that cause trypanosomiasis and chagas disease.

Many of these are "one-shot" disorders, which disappear for good from the affected patient if treated with an effective drug. The major pharmaceutical corporations—increasingly focused only on finding drugs for chronic illnesses that go on for years, and thus bring in long-term profits—have little incentive to find new drugs to fight such once easy-to-cure resistant disease organisms.

Along with bacteria, viruses, parasites and fungi, a number of larger creatures are also on track to become health threats as the world warms. These include, to give just two examples, Cuban tree frogs in the U.S. and oak processionary moth caterpillars in Britain.

The tree frogs, which likely arrived in Florida on imported palm trees, have been spreading their range, and are now found in New Orleans. "They have noxious skin secretions [irritating to humans], lay their eggs in bird baths and fish ponds, and can clog plumbing and cause power outages by short-circuiting power switches where they seek refuge," noted the US Geological Survey's Brad Glorioso.[27] The caterpillars, which "are covered in thousands of hairs containing the protein thaumetopoein, which can cause skin rashes, sore throats, breathing difficulties and eye problems" in people,[28] were accidentally introduced in Britain from continental Europe and are spreading northward. Those who touch them may experience rashes, vomiting and asthma attacks.

Though neither creature was introduced as a direct result of climate warming, climate may aid in extending their ranges, and their advent gives an idea of what can happen when any so-called "invasive species" is introduced—via climate warming or any other mechanism—to a new habitat.

The long history of species invaders, from rabbits and cane toads in Australia to European starlings in North America, is a mostly tragic one, and invaders riding a climate warming wave are likely to add many more negative chapters. One has only to speculate on the results when, for example, poisonous snakes like the coral snake, copperhead and cottonmouth start to extend their ranges from the southern U.S. northward into warming northern states and perhaps Canada. Bites from such critters are definitely not good for anyone's health.

Nor are potentially life-threatening stings from the bark scorpion, usually active only in the warmest months in the Sonoran Desert of Arizona, but enjoying a season-stretching boost from warmer temperatures. The scorpions came out early in 2016, thanks to rising temperatures, causing a surge in reported stings. According to researcher Israel Leinbach, their sting feels like "being stabbed with a hot knife," a feeling that can last for days. Because their venom is a neurotoxin, "nerves could fire uncontrollably. Muscles may spasm. Lips might twitch. Sometimes, eyes will roll around, in opposite directions."[29]

Even flying ants have joined the fray, by appearing ahead of schedule and disrupting the international pro-tennis tournaments at England's Wimbledon courts two years in a row. Though not a serious threat to players' health, they may have seriously affected their pocketbooks. "Flying ants made an unwelcome return to Wimbledon on Wednesday, prompting complaints from the reigning Australian Open champion Caroline Wozniacki," reported *The Guardian*.[30] "The bugs, which swarm in warm weather, caused problems for players last year when the British No. 1 Johanna Konta said she had swallowed some." The player, who suffered a surprise loss, complained: "They're in my mouth and in my hair and everywhere ... we need to do something. Is there a spray? I want to be here to focus on tennis, not eating bugs!"

"A Wimbledon spokeswoman said the bugs usually emerged at the end of the month but had appeared earlier this year [and last]."

Direct heat

Direct physical health threats from heat itself will also play an increasingly important part, as climate warming and its consequences continue. Sunstroke, heat prostration and even ordinary sunburn are likely to become more common. Of course, the temperature at which heat becomes a serious threat to human and animal health can vary considerably, depending on the humidity level at the time, and on what temperature ranges the people or animals are used to. A Maasai herder in western Kenya and his zebu cattle, for example, are acclimatized to much hotter, more humid days and nights than American or European dairy farmers and their Holsteins.

Since 1979, the U.S. National Weather Service has employed the so-called "heat index," which combines temperature and relative humidity to define "felt air temperature" in humans.[31] The U.S. National Oceanic and Atmospheric Association (NOAA) publishes a table showing the various combinations of temperature and humidity and the heat index values they produce.[32] Officially, a heat index of 105 F (41 C) or above is considered "dangerous," with heat stroke "likely," but much lower levels can—and have—proven fatal for many, particularly the elderly, or obese people, or those whose work requires them to be outdoors for longer periods.

According to the NOAA, an index value of 90 F (32 C) or above calls for "extreme caution: heat cramps and heat exhaustion are likely. Continuing activity could result in heat stroke." Heat stroke would come when a person's body temperature rises past 104 F (40 C). First, the skin reddens and the victim suffers headache, confusion and dizziness. If no relief is found, seizures, kidney failure and eventually death can follow.[33] Dehydration is another possibility, which when serious can cause headaches, confusion, unexplained tiredness, purple fingernails, seizures and, eventually death.[34]

Unknown unknowns

Of course, so many physical medical dangers, coming together in so short a time and from so many directions, are bound to create anxiety in their potential human victims, or as a 2011 article in *American Psychologist* put it more clinically, to create "threats to emotional well-being based on observation of impacts and concern or uncertainty about future risks."[45]

Will this lead to more people seeking relief from their fears via tranquilizers or addictive drugs?

Big Pharma, creators of the opioid crisis described in Chapter One, would doubtless react to that prospect with rapacious glee. The opioid epidemic has already killed more than 200,000 Americans.[46] In 2017 alone, it killed nearly 4,000 Canadians.[47] It has addicted hundreds of thousands more still-living victims, who continue to exist in misery. But it has been a gold mine for drug makers. Just one drug, OxyContin, has returned more than $31 billion in profits to the company that introduced it.[48]

The easy availability online of some potentially addictive drugs, otherwise obtainable only via prescription, exacerbates the danger. One London clinic reports an increase in both teenagers and adults addicted to such drugs, bought illegally on the Web. "Teenagers tend to use [prescription drugs] for the intoxicating effect, to get giddy and drunk, but older people tend to use it to treat symptoms, particularly anxiety," said Dr. Owen Bowden.[49]

Finally, there are what an American politician once confusingly dubbed "the unknown unknowns,"[50] which may come as surprise threats to health. Some, like the opioid crisis, may be deliberately manufactured by companies supposedly devoted to protecting our well-being. But others may pop up totally out of the blue, catching everyone off guard and unprepared.

Stay tuned ...

Food Supply

Next to health, securing adequate, nourishing food is probably our most serious concern. The coming multi-system collapse will make this far more difficult, and in some cases impossible.

As already noted in Chapter Two, the most immediately dangerous threat to agriculture, especially in Europe and North America, is probably the destruction of our grossly under-appreciated allies, the pollinators. All flowering plants, or angiosperms, require pollination to reproduce, and of the roughly 250,000 species of angiosperms, only a minority are capable of self-pollination. Most—87.5 per cent—require the help of outside vectors, chiefly insects, birds and bats, and in some cases the wind.

At least "3/4 of the world's food supply are plants that require pollinators,"[51] and much of the rest of that supply, such as the livestock and poultry that provide meat, milk and eggs, require flowering plants, such as alfalfa, for feed. Self pollinating crops such as corn (maize) or potatoes are in a definite minority, and insufficient in themselves to provide all of the key micronutrients, the vitamins, minerals and other substances, needed to maintain health. Even if such crops could supply more nutrients than they do, they too are in danger from climate change.

A recent study by British researchers, for example, found that extreme weather, particularly heat and drought, could threaten world maize production. The study "envisaged simultaneous catastrophic disruptions in China and the U.S. In 2014, total world production of maize was around one billion tons, with the U.S. producing 360 million tons and China growing 215 million. If production in these two countries were hit by simultaneous extreme weather events, most likely droughts, more than 60 per cent of global maize production would be hit.

"A double whammy like this has never happened in the past, but the work by the Met Office indicates there is now a real risk."[52]

The combined forces of factory farming methods, plus climate change, may wipe out most or all of the planet's living pollinators, or at least reduce their numbers so drastically that normal crop and livestock production will become a hopeless endeavour.

Factory farming, as already demonstrated in Chapter Two, has proven devastating to pollinators, destroying the habitats of birds and bats, and dosing crops with such massive amounts of pesticides and herbicides that it is a miracle any insects at all remain alive in our industrial farming regions. The toll of glyphosates and neonicotinoids has been particularly heavy on bees, both native wild bees and European honeybees. Butterflies, night-flying moths and other species have also been hit by chemicals intended to protect crops.

Even insect diseases, such as the honeybee-killing American Foulbrood disease, may have become resistant to antibiotic treatment as a result of the presence of an antibiotic-resistant gene in genetically manipulated Roundup-Ready crops. The gene makes bacteria resistant to tetracycline, used to treat AFB. [Roundup, as noted in Chapter Two, has also been implicated as a possible human carcinogen, a danger its makers have been accused in several lawsuits of covering up.][53]

Honeybee numbers have dropped drastically, especially in the U.S., where, as noted in Chapter Two, one in three honeybee colonies died over the winter in 2013. Wild bees have also taken a hit. In March 2017, the rusty patched bumblebee (*Bombus affinus*) "became the first bee in the lower 48 states to be listed as an endangered species after its population shrank by an estimated 91 per cent in 20 years. Other bumble bees, such as the American bumble bee (*B. Pennsylvanicus*) have declined more gradually."[54]

As noted earlier, in the U.S., commercial beekeepers have reacted to local pollinator shortages by trucking bee colonies from farm to farm, "following the bloom from south to north."[55] Unfortunately, this has resulted in the rapid spread of any diseases carried by the transported bees—such as resistant strains of American Foulbrood—to new areas, where they may infect previously unaffected insects. The cure ends up being as bad, literally, as the disease. A similar mechanism is at work via efforts to curb the spread of West Nile carrying mosquitoes, by aerial spraying of pesticides. The spray kills local pollinating insects, as well as mosquitoes.

Overall, insects everywhere are in the midst of an unprecedented population decline. A new report from Germany, for example, "has shown a 75 per cent plunge in insect populations since 1989, suggesting that they may be even more imperilled than any previous studies suggested."[56] According to another study, one-fifth of Europe's wood beetles, a key source of food for the small mammals, bats and birds who pollinate plants "are at risk of extinction."[57] The beetles, some species of which are themselves pollinators, are dying out due to a "widespread decline in ancient trees."

Now climate change is adding to the stress on insect populations. "We know that many insects are in rapid decline due to factors such as habitat loss and intensive farming methods," said Prof. Dave Goulson of the University of Sussex, UK.[58] He noted that a recent study "shows that, in future, these declines would be hugely accelerated by the impacts of climate change, under realistic climate projections." The study he cited, at the University of East Anglia, showed that **typical insect species "would likely lose 43 per cent of their range" if climate warms as expected by the year 2100**.

As Harvard University entomologist E.O. Wilson has warned, insects are to terrestrial food chains what plankton are to oceanic ones. "Without insects and other land-based arthropods, he estimates, humanity would last all of a few months. After that, most of the amphibians, reptiles, birds and mammals would go, along with the flowering plants. The planet would become an immense compost heap, covered in shoals of carcasses and dead trees that refused to rot ... The earth would revert to what it was like in the Silurian period, 440 million years ago, when life was just beginning to colonize the soil."[59]

Birds too are suffering, partly from the same causes that are destroying insects, but also because they depend so heavily on those same insects as a key source of food. Reported *The Observer's* science editor:

> *An insect Armageddon is under way, say many entomologists, the result of a multiple whammy of environmental impacts, pollution, habitat changes, overuse of pesticides, and global warming. And it is a decline that could have crucial consequences. Our creepy crawlies may have unsettling looks, but they lie at the foot of a wildlife food chain that makes them vitally important to the makeup and nature of the countryside ...*

> *The best illustration of the ecological importance of insects is provided by our birdlife. Without insects, hundreds of species face starvation and some ornithologists believe the lack of food is already causing serious declines in bird numbers. A point stressed by the naturalist and wildlife author Michael McCarthy. "Britain's farmland birds have more than halved in number since 1970," he points out. "Some declines have been catastrophic. The spotted flycatcher, a specialist predator of aerial insects, has declined by more than 95 per cent, while the red-backed shrike, which feeds on big beetles, became extinct in Britain in the 1990s."*[60]

In France, a dramatic fall in farmland birds such as skylarks, whitethroats and ortolan buntings was revealed by two studies, with the spread of neonicotinoid pesticides and decimation of insect life coming under particular scrutiny ... As noted in the previous chapter, French farmland bird numbers have dropped by a third in 15 years.[61]

Birds—for example, the ruby throated hummingbird of North America—and insects alike are pollinators, and thus crucial to agricultural crop production. Birds also help to control many insect pests, that eat crops or attack livestock.

~

Everything is connected to everything else. Warming expands the ranges of noxious insects, which destroy crops or infect people with disease, at the same time that it curtails the ranges of valuable pollinating insects, or prompts pesticide spraying that kills off local pollinators. It alters the ranges of birds, or their reproductive patterns, or their food sources.

It may also extend the ranges of unwanted plants, like the invasive "superweed" Palmer Amaranth (*Amaranthus palmeri*), commonly called "pigweed," mentioned in Chapter Two [and not to be confused with "giant hogweed," *Heracleum mantegazzianum*]. The monster plant, which can grow as fast as three inches a day and reach heights of eight feet, was originally native to the southwestern U.S., but has spread in recent years to 28 U.S. states, as far north as Minnesota, infesting cotton and soybean fields. Part of the spread has been due to accidental transport of seed in animal feed or manure, but climate models call for it to spread even further, as far as North Dakota, as warming habitats increase.

The weed has become resistant to several herbicides.

And of course, climate change may physically destroy crops or livestock, via drought, heat waves or flash floods. Studies in the *Proceedings of the National Academy of Sciences* have warned human-caused climate disruption, particularly drought, will "increase the risk of simultaneous crop failures across the planet's largest corn-growing regions, as well as sapping nutrients from critical vegetables. For example, an increase of four degrees C, which is essentially the current trajectory we are on to reach by 2100, could cut U.S. corn production nearly in half."[62]

California's Central Valley, which produces more than a quarter of the [U.S.] nation's fruits, nuts and vegetables, could by the end of the century experience "temperatures rivalling Death Valley's and face the loss of 90 per cent of the Sierra Nevada snowpack, the region's main water source."[63] Research at the University of California-Davis shows "higher temperatures will likely decimate the state's $10 billion fruit and nut industry."[64]

Wind erosion, caused by drought, also blows away topsoil, which—due to the combined effects of heat stress, and to the factory farm soil cultivation regimes and monocropping described in Chapter Two—may already have lost many of the soil organisms that once enriched it and provided food for crop roots.

That microbial diversity, the soil biome, may also have been affected, or could be affected in future, by climate warming. Just as climate change may be increasing the ranges of disease-causing microbes, it can alter the ranges and populations of beneficial soil microbial communities. At the same time, those communities may themselves be affecting climate in ways we don't yet understand. Some of the interactions of microbes and climate are extremely complex, and even counter-intuitive.

For example, Steven Schmidt of the University of Colorado studies microbial activity under the snowpack in coniferous forests. Microbial activity is especially high in late winter, when the temperature is optimum for the growth of snow moulds, which produce the greenhouse gas CO_2. "As the soils warm [due to climate change], snow moulds will have less water and will produce less CO_2, which may sound good in terms of global warming, but the trees in this system [which take in CO_2 and give off oxygen] also depend on snow melt water and will ultimately die under extreme drought, thus leading to an overall decrease in formation of carbon [which the trees need] by the system," Schmidt explains. "The trees may die. Overall, it's probably going to be bad."

The situation is complicated by the fact that, while snow moulds give off CO_2, other microbes—particularly mycorrhizal fungi—absorb it. The fungi exude a compound called glomalin, a glue-like glycoprotein that literally "gloms onto" carbon, holding it in the soil. A recent University of Maryland study found that glomalin "accounts for 27 per cent of the carbon in soil."[65] Another study, reported in *Nature*, found that enhancing soil's ability to sequester carbon "could reduce greenhouse gas concentrations by between 50 and 80 per cent."[66]

Further complicating these relationships are the various other functions and potential functions, that soil microbes perform. "Plants treated with soil microbes have a deeper root system and their shoots grow more quickly. Consequently, under drought stress, plants inoculated with microbes can more effectively take up water from drying soil."[67] And again, "soil microbes applied directly to seeds helped plants combat the rice leaf-folder insect, an important rice pest."

Thus, both climate warming and farming practices that destroy the soil biome threaten agriculture, but in ways that may not be immediately apparent. Nor is it always immediately apparent how crucially valuable certain soil microbes might be, or how dangerous if their populations are thrown seriously out of balance. Here is another place where the Precautionary Principle ought to apply. Only fools rush in ...

And finally, factory farming, through its dependence on a small number of high-yield, often hybrid crop varieties, or on genetically modified (GMO) varieties, reduces the genetic resources available from older conventional crops. Those lost or forgotten conventional varieties may have had inbred tolerances for hotter, drier weather, traits we might need today to contend with climate warming.

Like lost insect species or bird species, they might even have carried, "inside their exotic biochemistries, cures for any number of diseases. Recently, chemicals harvested from sea slugs have been tested in clinical trials in the U.S. for use as cancer fighting drugs. Others could be used as natural alternatives to pesticides ..." It is, indeed, "a kind of vandalism."[68]

The Economy

Climate change will also have a pronounced effect on the economic and social organization of human societies, at the same time that it is affected and often exacerbated by the economic and political games people play. In this respect, perhaps more obviously than any other, people risk "shooting themselves in the foot." The factors cited in Chapter Three are all likely to come into play, directly or indirectly.

That chapter focused first on the economic ills already besetting younger people—the so-called Millennials and those born after them, ills which make them more vulnerable to and less able to ride out future society-wide financial disasters. It also discussed the factors that have caused major economic catastrophes in the past, such as the Great Depression of the 1930s and the so-called Great Recession of 2008.

One of the five chief factors commonly listed as causes of or influences worsening the 1930s debacle was the **severe drought in the so-called "Dustbowl"** states. As already noted, climate warming could bring about "the Mother of All Dustbowls," not only in the U.S., but across large portions of the world, particularly in the global south. And just as the original American version produced the "Okies," destitute farm families driven off their land by the heat, so similar future catastrophes, multiplying across the warmer regions of the earth, will produce new, and far larger, waves of climate refugees.

First, their farms and their food output will be lost to their home country economies and to international trade, then the farm owners and workers, joining the ranks of the unemployed, will create a further

Security Project, said: "Climate change could lead to a humanitarian crisis of epic proportions. We're already seeing migration of large numbers of people around the world because of food scarcity, water insecurity and extreme weather, and this is set to become the new normal."

"Climate change impacts are also acting as an accelerant of instability in parts of the world on Europe's doorstep, including the Middle East and Africa," Cheney said.[72]

Northern countries should not assume that they will be spared the kinds of natural shocks driving people out of the global south. Farmers in Britain, for example, were facing "tinderbox conditions" in 2018, which "severely reduced grass growth and depleted yields for many crops, leading to concerns that there will be a shortage of feed for livestock and dairy farmers later in the year ... The heat wave has acted as a 'timely reminder' that the UK cannot take its food production for granted, according to National Farmers Union President Minette Batters."[73]

Parts of England "had no rain for more than 50 days, and the Agricultural and Horticultural Development Board has said it is the driest runup to harvest in 80 years." Prices were directly affected, with wholesale prices for lettuce jumping up 22 per cent, and for carrots and onions 55 per cent.[74]

In addition, "wildfire is now an overriding concern for many farmers. On land this dry, with weather this hot, fires can catch quickly and spread fast ... Crops will be particularly vulnerable as they are ripening at present, drying out under the sun."[75]

And, as noted in Chapter Three, all of the other non-climate-caused factors that provoke economic crises are also in place today, ready to be aggravated by the added stress of a warming world.

A **reduction in purchasing power** is already underway, first for the heavily-indebted and under-employed young, struggling to manage student loan debts while facing few opportunities for full-time, decently paid employment. According to a 2018 report by the Brookings Institution, student loans are the only form of consumer debt that has continued to grow in the wake of the housing crisis that led to the 2008 recession. The report found that the current U.S. student debt of $1.3 trillion appears to be heading towards a similar crisis, and that "nearly 40 per cent of borrowers could default on their student loans by 2023."[76]

As for the population in general, including those young persons' elders, household debt is at dangerously high levels. In England, in 2018, for example, the figures were "worse than at any time on record," according to the Office for National Statistics. "The shortfall amounted to nearly 25 billion pounds, and the overspend was mostly paid for with borrowed money."[77]

High tariffs, leading to trade wars, begun with the Smoot-Hawley Tariff of 1930, preceding the Great Depression, have their counterpart today in the trade wars U.S. President Trump is busy launching with Europe and other major trading partners, such as Canada and China. In a 2018 economic prospects report, the World Bank warned that economic nationalism of the sort promoted by Trump could have "severe consequences" for world trade and economic growth, "similar to the drop experienced during the financial crisis of 2008-09."[78]

Meanwhile, the risk of **failures of banks and other financial institutions**, a major cause of the Great Depression and the Recession of 2008, has increased as the Trump administration "unwinds the post-crisis regulations in the banking sector," including the Dodd-Frank Act, and weakens government regulatory agencies such as the Consumer Financial Protection Bureau.[79] The rollback of the Dodd-Frank Act makes it "now very likely that the 'toxic, speculative activities' of the Wall Street crowd will return with a menace."[80]

Former British Prime Minister Gordon Brown recently warned that "we are in danger of sleepwalking into a future [economic] crisis,"[81] but that this one could be far worse than the 2008-9 recession because, unlike in the earlier crisis, "a breakdown of trust in the financial sector would be mirrored by breakdown in trust between governments. There wouldn't be the same willingness to cooperate but rather a tendency to blame each other for what's going wrong." This he attributed to "Trump's protectionism."

Finally, as described in Chapter Three, **stock market crashes, often called "panics,"** are partly prompted by "crowd behaviour," and might other factors. One is that ... improvements in education and literacy is making far more people aware of their own marginalisation and unwilling to accept it, and the other that there is an assumption in the West that security can best be assured by resort to military responses ... At the crudest level, what is sometimes termed the 'control paradigm' might better be termed 'liddism': keeping the lid on problems rather than understanding their causes and manifestations."[93]

The West failed, and still fails to see that "marginalisation and climate disruption would lead to a thoroughly unstable world that could not be controlled by minority elites, however powerful their military forces ... the free market is seen as the only way forward, but since it cannot handle climate change, then climate change cannot be happening."[94] And the threat is no longer exclusively from without. "What we are talking about here is not a minority fringe that is consistently excluded from the benefits accruing to most of society, but the majority of people, who are sharing inadequately the benefits of distorted economic growth."[95]

Efforts to contain the restless millions of the marginalised abroad have continued, morphing from "boots on the ground" to reliance on unmanned drones to deliver punishment. From 2009 to 2012, "the U.S. mounted 1,160 armed-drone attacks in Afghanistan alone." Also, from 2010 to 2015, many commercial developments in the area of public order control, including anti-riot agents, have become available in forms that may be deliverable by drone."[96]

Or by militarized local police, outfitted not just with water cannons and tear-gas, but with armoured cars and tanks, as such civilian police forces, everywhere, increasingly are.

In future, when waves of desperate environmental refugees flooding in from abroad are joined by rising numbers of disaffected, distressed native-born citizens in the rich countries, how will they be contained? By drones, raining death from above not only on people in the Middle East or Africa, but on the rebellious citizens of the U.S., Canada and Europe, in their own cities? Will we witness the sort of dystopian, drone-filled world envisioned in science fiction movies starring Arnold as *The Terminator*? Would Trump, or a future Trump clone, go so far as to order a nuclear strike on a rebellious American city, rather than lose his grip on power? Would the military permit it?

Should such extreme fears, of an extreme world, even be taken seriously?

There are indications that they are being taken seriously by the U.S. Army, whose Strategic Studies Institute recently warned that a series of domestic crises could initiate large-scale civil unrest. "Under the most extreme circumstances," Department of Defense (DOD) action "might include use of military force against hostile groups inside the United States. Further, DOD would be, by necessity, an essential enabling hub for the continuity of political authority in a multi-state or nationwide civil conflict or disturbance."[97] The chief of the U.S. "Joint and Army Concepts Division, in 2010, described preparations for fighting in the homeland as a way to legitimize the U.S. military budget."[98]

Many in the military have come to see endless war against "terrorist" or simply rebel groups as a form of "victory,"[99] based on the contralogical idea that if you are still fighting you aren't losing, and thus you are winning. Such an endless "victory" could occur anywhere, including a rebellious U.S.

A pretty strong indication of our real danger, not only in the U.S. but worldwide, came recently from a source which has, too often, been the caboose on any socially progressive train: the Roman Catholic hierarchy. At a Vatican conference marking the third anniversary of his 2015 environmental encyclical, *Praise Be*, Pope Francis warned that if climate change, unsustainable development and rampant consumption continue unabated, "there is a real danger that we will leave future generations only rubble, deserts and refuse."[100]

Is it already too late, as many in the scientific community fear? Or is there some faint hope still of mitigating, or as a species at least surviving what has all the earmarks of becoming a genuine, real-time Apocalypse?

If there *is* still hope, what can we do? And how can we, how *should* we do it?

Let's start with what we probably shouldn't do ...

---

1 Survival Condo, "Why a survival condo" February 2017, as posted online 20 June 2017 at http://survivalcondo.com/in-depth/, 3.
2 Daisy Luther, "Survival condos: a tour of bunkers for rich people in the middle of Kansas," 9April 2018, *The Organic Prepper*, as posted online 30 April 2018 at www.theorganicprepper.com/survival-condos/, 3.
3 Op. Cit., Survival Condo, 3-4.
4 *The Gathering*, "Leaving on a jet plane," 1 February 2017, as posted online 20 June 2017 at https://thegathering.com/jetplane/, 2.
5 Doublas Rushkoff, "How tech's richest plan to save themselves after the apocalypse," 24 July 2018, *The Guardian*, as posted online 26 July 2018 at www.theguardian.com/technology/2018/jul/23/tech-industry-w ..., 1-2.
6 John Steinbeck, *The Grapes of Wrath*, (New York: Viking Penguin, 1986).
7 Draper L. Kauffman Jr., *Systems 1: an introduction to systems thinking*, (Minneapolis, Minn.: S.A. Carlton, Publisher, 1980), 38.
8 *Wikipedia, the Free Encyclopedia*, "List of antibiotic resistant bacteria," as posted online 7 May 2018 at https.://en.wikipedia.org/wiki/List_of_antibiotic_resistant_bacteria, 1-5.
9 *Op. Cit.*, 1.
10 *Centers for Disease Control and Prevention*, "Biggest threats: antibiotic/antimicrobial resistance," 27 February 2018, as posted online 7 May 2018 at www.cdc.gov/drugresistance/biggest_threats.html
11 Caroline Plante, "In the race to fight antibiotic resistance, the livestock industry can be a game changer," 12 January 2017, *STAT*, as posted online 12 May 2018 at www.statenews.com/2017/01/12/antibiotic-resistance-livestock- ..., 2.
12 Slimdoggy, "Slimdoggy health check: bacterial diseases in dogs," as posted online 12 May 2018 at http://slimdoggy.com/slimdoggy-health-check-bacterial-diseases-in-dogs/, 2.
13 Nick Perry, "New Zealand to slaughter 150,000 cows to end strain of disease-causing bacteria," 28 May 2018, *The Associated Press*, as posted online 28 May 2018 at www.thestar.com/news/world/2018/05/28/new-zealand-to-slau ...
14 Barbara D. Petty, DVM, Ruth Francis-Floyd, DVM, "Bacterial diseases of fish," 2018 *Merck Veterinary Manual*, as posted online 12 May 2018 at www.merckvetmanual.com/exotic-and-laboratory-animals/aqu ..., 3.
15 Plante, *Op. Cit.*, 1.
16 *Loc. Cit.*
17 Press Association, "Antibiotic resistance could spell end of modern medicine, says chief medic," 13 October 2017, *The Guardian*, as posted online 14 October 2017 at www.theguardian.com/society/2017/oct/13/antibiotic-resistanc ..., 1.
18 Robin McKie, "'Antibiotic apocalypse:' doctors sound alarm over drug resistance," 8 October 2017, *The Guardian*, as posted online 8 October 2017 at www.theguardian.com/society/2017/oct/08/world-faces-antibio ..., 1-2.
19 Dr, Ken Tudor, "New hope for antibiotic resistance in humans and pets," 24 February 2015, *Pet MD*, as posted online 12 May 2018 at www.petmd.com/blogs/thedailyvet/ken-tudor/2015/february/n ..., 1.
20 Leah Cowen, "Infectious diseases are poised for a deadly comeback," 10 February 2018, *The Toronto Star*, as posted online 10 February 2018 at www.thestar.com/opinion/contributors/2018/02/10/infectious-d ..., 2.
21 Nicola Davis, "Antibiotic resistance crisis worsening because of collapse in supply," 31 May 2018, *The Guardian*, as posted online 31 May 2018 at www.theguardian.com/society/2018/may/31/antibiotic-resistan ..., 1-2.
22 Caroline Chen, "FDA repays industry by rushing risky drugs to market," 30 June 2018, *Truthout, Pro Publica*, as posted online30 June 2018 at https://truthout.org/articles/fda-repays-industry-by-rushing-risky-drugs-to-market/

23 *Wikipedia, the Free Encyclopedia*, "Lyme Disease," as posted online 3 June 2018 at https://en.wikipedia.org/wiki/Lyme_disease#North_America,
24 Janet Ioehrke and Karl Gelles, "Diseases on the move because of climate change," 5 December 2013, *USA Today*, as posted online 2 June 2018 at https://www.usatoday.com/story/news/nation/2013/12/04/climate-cha ..., 3.
25 Rene Cho, "How climate change is exacerbating the spread of disease," 4 September 2014, *State of the Planet*, Earth Institue, Columbia University, as posted online 2 June 2018 at http://blogs.ei.columbia.edu/2014/09/04/how-climate-change-is-exace ..., 3.
26 *Wikipedia, the Free Encyclopedia*, "Antimicrobial resistance," as posted online 19 May 2018 at https://en.wikipedia.org/wiki/Antimicrobial_resistance, 7.
27 The Associated Press, "Invasive, fist-sized Cuban treefrogs discovered in New Orleans," 1 May 2018, *The Guardian*, as posted online 1 May 2018 at www.theguardian.com/environment/2018/may/01/new-orleans ...
28 The Associated Press, "Toxic caterpillars causing people to vomit and have asthma attacks are taking over London," 29 April 2018, *The Toronto Star*, as posted online 29 April 2018 at www.thestar.com/news/world/2018/04/29/toxic-caterpillars-ca ...
29 Fernanda Santos, "Lured by early warm weather, scorpions emerge to swarm Arizona homes," 28 April 2016, *The New York Times*, as posted online 15 July 2018 at www.nytimes.com/2016/04/29/us/lured-by-early-warm-weathe ..., 1-3.
30 Haroon Siddique, "Flying ants cause chaos during Wimbledon invasion," 4 July 2018, *The Guardian*, as posted online 4 July 2018 at www.theguardian.com/sport/2018/jul/04/flying-ants-cause-cha ...
31 *Wikipedia, the Free Encyclopedia*, "Heat index," as posted online 2 July 2018 at https://en.wikipedia.org/wiki/Heat_index, 1.
32 *Op. Cit.*, 3.
33 *Loc. Cit.*
34 *Wikipedia, the Free Encyclopedia*, "Dehydration," as posted online 2 July 2018 at https://en.wikipedia.org/wiki/Dehydration, 1.
35 Katherine Harmon, "How does a heat wave affect the human body?" 23 June 2010, *Scientific American*, as posted online 2 July 2018 at https://www.scientificamerican.com/article/heat-wave-health/
36 Dennis Thompson, "Expect more deadly heat from climate change, study suggests," 27 March2017, *HealthDay*, as posted online 26 June 2018 at https://consumer.healthday.com/diseases-and-conditions-information- ..., 1.
37 Mia Pattillo, "Study projects climate change to increase heat-related deaths," 13 September 2017, *Brown Daily Herald*, as posted online 29 July 2018 at www.browndailyherald.com/2017/09/13/study-projects-climate ...
38 Press Association, "Heat-related deaths will rise 257% by 2050 because of climate change," 4 February 2014, *The Guardian*, as posted online 29 June 2018 at www.theguardian.com//environment/2014/feb/04/heat-related-d ..., 2.
39 Somim Sengupta, "In India, summer heat may soon be literally unbearable," 17 July 2018, the *New York Times*, as posted online 17 July 2018 at www.nytimes.com/2018/07/17/climate/ingia-heat-wave-summer.html
40 Daniel Hurst, "Japan heatwave: record broken as concern grows over 2020 Olympics," 23 July 2018, *The Guardian*, as posted online 23 July 2018 at www.theguardian.com/world/2018/jul/23/japan-heatwave-pro ...
41 Justin McCurry, "Japan: death toll from record rain increases as PM warns of 'race aganst time.'" 8 July 2018, *The Guardian*, as posted online 8 July 2018 at www.theguardian.com/world/2018/jul/08/death-toll-increases-a ...
42 Oliver Milman, "Wildfire smoke: experts warn of 'serious health effects' accross western U.S.," 2 August 2018, *The Guardian*, as posted online 2 August 2018 at www.theguardian.com/world/2018/aug/02/wildfire-events-air- ..., 1.
43 Denis Campbell and Sarah Marsh, "Upsurge in sleeping problems due to UK's

longest heatwave in 40 years," 13 July 2018, *The Guardian*, as posted online 13 July 2018 at www.theguardian.com/science/2018/jul/13/upsurge-in-sleeping ...

44 Matthew Weaver, "Heatwave forces UK farmers into desperate measures to save cattle," 28 June 2018, *The Guardian*, as posted online 28 June 2018 at www.theguardian.com/uk-news/2018/jun/28/heatwave-forces- ..., 1.

45 *Wikipedia, the Free Encyclopedia*, "Effects of global warming on humans," as posted online 16 December 2017 at https://en.wikipedia.org/wki/Effects_of_global_warming_on_humans, 2.

46 Nicole Colson, ""Will pharmaceutical companies ever be punished for the opioid crisis?" 28 April 2018, *Truthout*, as posted online 28 April 2018 at www.truth-out.org/opinion/item/44304-will-the-real-opioid-pus ..., 2.

47 Karen Howlett, "Ottawa urges drug makers to end opioid marketing," 27 June 2018, *The Globe & Mail*, as posted online 27 June 2018 at www.theglobeandmail.com/canada/article-ottawa-asks-drug-companies-to-stop-marketing, 1.

48 *Colson, Truthout, Op. Cit., 2.*

49 Sarah Marsh, "Rise in people seeking help over prescription pills bought online," 13 July2018, *The Guardian*, as posted online 13 July 2018 at www.theguardian.com/society/2018/jul/13/doctors-warn-of-ris ...

50 *Wikiquote*, "Donald Rumsfeld," as posted online 26 June 2018 at http://en.wikiquote.org/wiki/Donald_Rumsfeld

51 *Wikipedia, the Free Encyclopedia*, "Pollination," as posted online 7 July 2018 at https://en.wikipedia.org/Pollination

52 Robin McKie, "Maize, rice, wheat: alarm at rising climate risk to vital crops," 15 July 2017, *The Guardian*, as posted online 15 July 2017 at www.theguardian.com/environment/2017/jul/15/climate-chang ..., 2.

53 Carey Gillam, "Landmark lawsuit claims Monsanto hid cancer danger of weedkiller for decades," 22 May 2018, *The Guardian*, as posted online 22 May 2018 at www.theguardian.com/business/2018/may/22/monsanto-trial-c ...

54 Kelsey K. Graham, "Beyond honey bees: wild bees are also key pollinators, and some species are disappearing," 3 June 2018, *The Conversation, Truthout*, as posted online 3 June 2018 at https://truthout.org/articles/beyond-honey-bees-wild-bees-are-also-key-pollinators-and-some-species-are-disappearing/

55 Op. Cit., *Wikipedia*, "Pollination," 6.

56 Jacob Mikanowski, "A different dimension of loss: inside the great insect die-off," 14 December 2017, *The Guardian*, as posted online 3 May 2018 at www.theguardian.com/environment/2017/dec/14/a-different-dimension-of-loss-great-insect-die-off-sixth-extinction ...

57 Arthur Neslen, "One-fifth of Europe's wood beetles at risk of extinction as ancient trees decline," 5 March 2018, *The Guardian*, as posted online 5 March 2018 at www.theguardian.com/environment/2018/mar/05/a-fifth-of-eu ...

58 Damian Carrington, "Climate change on track to case major insect wipeout, scientists warn," 17 May 2018, *The Guardian*, as posted online 17 May 2018 at www.theguardian.com/environment/2018/may/17/climate-cha ..., 2.

59 *Op. Cit.*, Mikanowski, 4.

60 Robin McKie, "Where have all our insects gone?" 17 June 2018, *The Guardian*, as posted online 17 June 2018 at www.theguardian.com/environment/2018/jun/17/where-have-i ..., 2.

61 Patrick Barkham, "Europe faces 'biodiversity oblivion' after collapse in French birds, experts warn," 21 March 2018, *The Guardian*, as posted online 21 March 2018 at www.theguardian.com/environment/2018/mar/21/europe-faces ...

62 Dahr Jamail, "Global temperatures projections could double as the world burns," 16 July 2018, *Truthout*, as posted online 16 July 2018 at https://truthout.org/articles/global-temperature-projections-could-double-as-the-world-burns/

63 Josh Harkinson, "The new dust bowl," November 2009, *Mother Jones*, as posted

online 27 January 2017 at www.motherjones.com/environment/2009/11/new-dust-bowl?page=2, 1.

64 *Ibid.*

65 United States Department of Agriculture, "Glomalin: hiding place for a third of the world's stored soil carbon," September 2002, *AgResearch Magazine*, as posted online 23 July 2018 at https://agresearchmag.ars.usda.gov/2002/sep/soil, 1.

66 Esther Ngumbi, "How soil microbes fight climate change," 17 May 2016, *Scientific American Guestblog*, as posted online 23 July 2018 at https://blogs.scientificamerican.com/guest-blog/how-soil-microbes-fight-climate-change/

67 *Ibid.*

68 *Op. Çit.*, Mikanowski, 6.

69 Andrew Gumbel, "'They were laughing at us': immigrants tell of cruelty, illness and filth in US detention," 12 September 2018, *The Guardian*, as posted 12 September 2018 at www.theguardian.com/us-news/2018/sep/12/us-immigratio ..., 1.

70 Ruth Pollard, "The Mediterranean refugee crisis explained," 26 August 2015, *Sydney Morning Herald*, as posted online 26 August 2015 at www.smh.com.au/world/the-mediterranean-refugee-crisis-explained-20150826-gj7pgz.htm, 1.

71 Lorenzo Tondo, Angela Giuffrida, "Warning of 'dangerous acceleration' in attacks on immigrants in Italy," 3 August 2018, *The Guardian*, as posted online 3 August 2018 at www.theguardian.com/global/2018/aug/03/warning-of-dangero ...,1.

72 Damian Carrington, "Climate change will stir 'unimaginable' refugee crisis, says military," 1 December 2016, *The Guardian*, as posted online 4 July 2018 at www.theguardian.com/environment/2016/dec/01/climate-chang ..., 1.

73 Jamie Doward, "Farmers in drought summit amid fears of food supply crisis," 28 July 2018, *The Guardian*, as posted online 28 July 2018 at www.theguardian.com/environment/2018/jul/28/farmers-droug ..., 1.

74 Sarah Butler, "Heatwave pushes up UK fruit and vegetable prices as yields fall," 27 July 2018, *The Guardian*, as posted online 27 July 2018 at www.theguardian.com/business/2018/jul/27/heatwave-pushes- ..., 1.

75 Fiona Harvey, "British farmers fear fire as heatwave creates 'tinderbox'" 25 July 2018, *The Guardian*, as posted online 25 July 2018 at www.theguardian.com/environment/2018/jul/25/british-farmers ...

76 Mona Chalabi, "Deceased and still in debt: the student loans that don't get forgiven," 1 August 2018, *The Guardian*, as posted online 1 August 2018 at www.theguardian.com/mney/datablog/2018/aug/01/deceased ...

77 Phillip Inman, "Household debt in UK 'worse than at any time on record,'" 26 July 2018, *The Guardian*, as posted online 26 July 2018 at www.theguardian.com/money/2018/jul/26/household-debt-in- ..., 1.

78 Richard Partington, "World Bank warns trade tensions could cause 2008-level crisis," 5 June 2018, *The Guardian*, as posted online 5 June 2018 at www.theguardian.com/business/2018/jun/05/world-bank-warn ...

79 David Olive, "Trump is unwinding bank regulations. Here's why Canadians should worry," 27 April 2018, *Toronto Star*, as posted online 27 April 2018 at www.thestar.com/business/opinion/2018/04/27/trump-is-unwin ...

80 C.J. Polychroniou, "Goodbye regulations, hello impending global financial crisis," 3 July 2018, *Truthout*, as posted online 3 July 2018 at https://truthout.org/articles/goodbye-regulations-hello-impending-global-financial-crisis/, 1.

81 Larry Elliott, "The world is sleepwalking into a financial crisis' – Gordon Brown," 12 September 2018, *The Guardian*, as posted online 12 Septem ber 2018 at www.theguardian.com/politics/2018/sep/12/we-are-in-dange ..., 1-2.

82 Chris Field, "What are the economic consequences of climate change?" 16 November 2015, *World Economic Forum*, as posted online 25 July 2018 at www.weforum.org/agenda/2015/11/what-are-the-economic-co ...

83 Justin Worland, "Climate change could wreck the global economy," 22 October

After checking in with the International Committee for the Red Cross (ICRC) people, we drove downtown, stopping near the city centre and getting out to walk. It was a walk through a charnel house. Everywhere, dead bodies littered the ground, wrapped, according to Somali burial custom, in bolts of dingy cloth with a sort of knot at the top, like macabre, oversized imitations of fancy wrapped toffees. Living people, themselves looking like walking skeletons, stepped over the bodies.

From time to time, someone in the crowd would start to stagger, like a drunk, then fall flat in the dirt—exhausted by malnutrition. They would lie there, gasping out their last few breaths, feebly moving their arms, then die. People passed carrying homemade stretchers piled with dead bodies, and stepped over the dying ones. In the background, bursts of small arms fire sounded, like strings of firecrackers popping. No one reacted. Our guards grinned broadly, chuckling.

It was incredible.

We spotted a woman sitting on the ground in front of a roofless, windowless house (the tin roof had long since been carried off and sold for scrap, the windows smashed). Next to her was a small, cloth-wrapped package, about the size of a child. The woman was crying. "What is in that bundle?" the NPR man asked her, holding his microphone towards her for a response. Our interpreter translated into Somali, and the woman, at first uncomprehending, then answering frankly—perhaps relieved to have someone to whom she could pour out her grief—said it was her child. Dead of starvation.

I turned away. The NPR man was only doing his job, recording an interview that might prompt his eventual listening audience to contribute hard cash for Somali relief. But I felt like a voyeur, watching the mother's pain. I walked a few steps away, and almost fell over another cloth bundle, this one with a bare, black foot sticking out the end. There was another, and another.

I looked in a different direction, and saw the whitewashed walls of another house with no windows. I walked up and looked inside, to see what might be there. Inside were four more bodies, also wrapped in dirty cloth. I turned back toward the street, in the opposite direction from the woman being interviewed by the NPR man. A man and a woman were walking down the road, carrying a square bedspring on which a dead body lay. A burst of small arms fire sounded suddenly at the end of the block, and I turned quickly to see. No one else turned. No one paid any attention.

Only if the bullets actually came close—a situation recognizable by the peculiar whistle/zipping sound high-power rounds make when they part the air—did people react. This happened at the crossroads near the only well left operating in town. One minute two long lines of people, some leading donkeys with blue water drums strapped to their backs, were waiting patiently for their turn at the water hose. The next, bullets whizzed by and dozens of men, women and children were scrambling for cover, ducking behind trees, stones, buildings, eyes round with fear, heads turning this way and that, trying to find out where the firing was coming from. They knew from painful experience when to duck and when not.

Precious water spilled from the hose onto the ground as drum-laden donkeys shied.

Only the day before, we were told, a similar burst had left a little girl of nine or 10 bleeding in the dirt, two bullets from an AK47 having passed through her stomach. Then it was the turn of a small boy, his arm taken nearly off at the elbow. This time, however, no one was hurt. The young gunmen several dozen yards away who fired the shots—it wasn't clear whether they were someone's militia or simply freelance bandits—meant nothing personal by it. High from chewing narcotic *khat* leaves, which cause a cocaine-like buzz, or maybe just horsing around as teenagers do, they were in the habit of shooting their weapons, as one local Red Cross worker put it, "pretty much in every direction." They likely wouldn't know, or even notice any victims.

And so the tour went, from worse to worst, dodging bullets, observing ruin, death and pain, narrowly missing becoming part of it when snipers' rounds whizzed by our own heads. It was so ugly, so outrageous, that I struggled to find words to describe it. It was a graphic demonstration of just how far human beings could degenerate, and their

Numerous reports have been published showing how drug companies paid doctors to promote dangerous painkillers to their patients. A 2003 government report revealed that, to promote OxyContin sales, "Purdue Pharma distributed several types of branded promotional items, to doctors and physicians, including OxyContin fishing hats, stuffed toys, luggage tags, music CDs, coffee mugs with heat activated messages. The company even created a program whereby doctors could distribute coupons to their patients for free, single-use OxyContin prescriptions."[7] Doctors received "lavish bonuses for their efforts—sometimes in the form of free food and beverages, and other times in the form of private stadium suites at sporting events and dinner at five-star restaurants."[8] Street-corner crack pushers should be so lucky.

The end result of all the hype was a trail of destruction even greater than that left by so-called organized crime:

> *"The National Institute on Drug Abuse reports that in 2001 around 2,000 people overdosed on heroin across the United States. By 2013 heroin claimed about 8,000 lives. As the trade and death count of heroin increased, so too did the number of deaths related to overdosing on prescription opioids, like Purdue Pharma's OxyContin. There were 6,000 such fatalities in 2001, by 2013 there were 15,000. The National Survey on Drug Use and Health discovered that 80 per cent of heroin users started out on opioids, dramatically changing the demographics of drug use across the country."*[9]

Thus, not only did the corporate product cause nearly three times more deaths than the illegal substance, it multiplied those caused by heroin, thus boosting the effects of actual organized crime activity. Today, there is an epidemic of babies being born already addicted to opioids, thanks to their opioid-addict mothers. Between 1999 and 2014, researchers found that the number of addicted mothers who gave birth to addicted infants "increased from 1.5 mothers per 1,000 deliveries to 6.5 per 1,000."[10]

In the U.S., the only real difference between a street pusher and some pharmaceutical salesmen, or doctors acting as sales boosters, seems to be the technical one of the corporate products not being listed by the Drug Enforcement Administration as a Schedule I or Schedule II substance, which would make them officially illegal. And the doctors wear suits.

Sometimes, rarely, corporate acts actually are deemed illegal, as in the 1999 case, noted in Chapter Four, where Philip Morris and other tobacco companies were found guilty of violating the U.S. Racketeer Influenced and Corrupt Organizations Act (RICO). In most such cases, however, the actual people whose decisions caused damage are not, as individuals, punished. The corporation, rather than its individual directors, is punished, by being ordered to perform some action or by being fined. And the fines, too often, represent less than major financial hardship, amounting to a wrist slap for an organization that may be worth billions.

For example, as noted in Chapter One, Purdue Pharma admitted that it "fraudulently and misleadingly marketed OxyContin as less addictive, less subject to abuse, and less likely to cause withdrawal symptoms than other pain medications," and eventually it paid a large fine for violating the rules. But the fine was small in comparison to the $30 billion in profits it had already made from the drug. [The company, which "is currently being sued by more than 1,000 jurisdictions for its alleged role in seeding the opioid crisis,"[11] denies the allegations in those suits.]

Still rarer are cases where corporate officers are themselves actually deemed criminals. In one 2019 case, a Boston jury found John Kapoor, the billionaire founder of Insys Theraputics, guilty of bribing doctors to prescribe a dangerous painkiller to patients who did not need it, as well as of defrauding insurance companies in a push to sell a spray "made from fentanyl, a synthetic opioid many times stronger than morphine."[12] Four other Insys executives were convicted on similar racketeering changes, and "faced up to 20 years in prison."

Such convictions, however, remain exceptions to the rule.

Meanwhile, at least one member of the family behind Purdue bodes well to keep right on profiting, this time, ironically, from marketing a medication to help opioid addicts. Former Purdue president Dr. Richard Sackler is one of six persons recently awarded a patent for a new version of buprenorphine,[13] which is used to treat addicts.

> *whole thing implodes. They steal hundreds of millions of dollars on Wall Street through fraud and theft, pay little or no taxes, almost never go to jail, write laws and regulations that legalize their crimes and then are asked to become trustees of elite universities and sit on corporate boards ... And if they get into legal trouble, they have high-priced lawyers and connections among the political elites to get them out ...*
>
> *The children of rich white families—surrounded by servants and coddled in private schools, never having to fly on commercial airlines or take public transportation—develop a lassitude, sometimes accompanied by a drug habit, that often leads them to idle away their lives as social parasites ... They are cultural philistines preoccupied with acquiring more wealth and more possessions. "Material success,' as C. Wright Mills wrote, "is their sole basis of authority."*

This author has mingled with them a few times, once at a Christmas holiday banquet in Switzerland, where a roast wild boar, with an apple actually in its mouth, was served on a silver platter (stirling), and another time at a cocktail party at a penthouse in Rome's exclusive Parioli quarter. I wore a uniform to the latter, a made-to-measure, hand-cut suit I'd once bought in Hong Kong (the only such suit I ever owned), expensive shoes, a Gucci necktie, bought for the occasion at the Gucci shop just down from the Spanish Steps, and the requisite Mont Blanc pen, with its distinctive snow-cap logo, tucked into my shirt pocket. On the surface, at least, it made me look like I was one of the bunch.

I sat on a sofa, while two men nearby played a game. One had just built a new *pied-à-terre* "outside Brasilia, in that section where we were last year, you know," and invited his interlocutor to "swing down and see us. I could send my new little Lear [meaning his private Lear jet aircraft] for you." The other man assumed a regretful look. "Oh no, couldn't this time of year. We're just finished fixing up the old place in the Alps, want to sort of break the chateau in again before anything else. Why don't you come visit us? My Lear is quite a bit bigger than yours; if I sent it you could bring friends."

The game, obviously, and pathetically, was one-up. My Lear is bigger than yours ... One didn't know whether to laugh or cry.

An exceptionally thorough, well written summary of the historical process that led to elite corporate control in the U.S. has been written by sociologist G. William Domhoff, of the University of California, Santa Cruz, who updates his 1967 book *Who Rules America?* regularly online.[19] Beginning in the U.S. year of independence 1776, he traces the growth of "rule by the wealthy few," maintaining that "members of this upper class control corporations, which have been the primary mechanisms for generating and holding wealth in the United States for upwards of 150 years."

He adds that, while "the upper class probably makes up only a few tenths of one per cent of the population, for research purposes I use the conservative estimate that it includes 0.5 to one per cent of the population." Hence the popular designation, prominent in recent years, of the "one per cent" versus everyone else.

Their domination of the corporate world, and corporate dominion over government, hasn't always been as complete as it is in the U.S. today, however. A long string of court cases, legislation and legal decisions by the U.S. Supreme Court has constantly expanded it, particularly those decisions imparting 1) "personhood" to corporations and 2) providing their directors and principals with so-called "corporate immunity." Domhoff, *Wikipedia* and other sources, discuss these at length.

In summary, legal **personhood**, which treats corporate organizations as if they were actual living humans, has developed through a series of decisions by the U.S. Supreme Court. Ironically, what has often been cited as the first of these was an error, in which a margin note written by a court reporter was taken as the actual court ruling on which he was commenting. That was the now-famous 1886 case of *Santa Clara County v. Southern Pacific Railroad Co.*[20]

However, a number of decisions before and since that incident have in fact established that "a corporation, separately from its associated human beings (like owners, managers or employees) has at least some of the legal rights and responsibilities enjoyed by natural persons (physical humans) ...

"The laws of the United States hold that a legal entity (like a corporation or non-profit organization) shall be treated under the law as a person except when otherwise noted. This rule of construction ... states: 'In determining the meaning of any Act of Congress, unless the context indicates otherwise, the words "person" and "whoever" include corporations, companies, associations, firms, partnerships, societies, and joint stock companies, as well as individuals.'

"This federal statute has many consequences. For example, a corporation is allowed to own property and enter contracts. It can also sue and be sued and held liable under both civil and criminal law."[21] Also among the consequences of this personhood is the conferring of **immunity**. "Because the corporation is legally considered 'the person' **individual shareholders are not legally responsible for the corporation's debts and damages beyond their investment in the corporation. Similarly, individual employees, managers and directors are liable for their own malfeasance or lawbreaking while acting on behalf of the corporation, but are not generally liable for the corporation's actions.**"[22]

As a result of this, it is often difficult, if not impossible, to seek damages from or criminally prosecute individuals, such as corporate directors or employees, for crimes deemed committed by the corporation. They can, to make a rough analogy, claim "the gun shot him, not me." The "gun" being legally a person.

In addition, the Supreme Court's 2010 decision in *Citizens United v. Federal Election Commission*, ruled that corporations, as persons, enjoyed the same free speech rights as physical humans under the U.S. Constitution. This, and a series of subsequent rulings by both the Supreme Court itself and federal courts of appeal, has resulted in a campaign finance regime where the elite and corporations can, almost literally, buy elections, particularly via so-called Super PACs (for Political Action Committees).

Ordinary PACs, which can solicit money from a business's employees or a union's members, have been around since the 1940s. But both the amount they can solicit and the amount they can subsequently donate to candidates or political parties has been limited fairly strictly, in both the U.S. and other countries. Not so Super PACs.

The *Citizens United* Supreme Court decision, and another ruling only two months later by the federal Court of Appeals for the District of Columbia (Washington, D.C.), in *Speechnow.org v. FEC,* allowed their creation. In *Speechnow,* the court "held that PACs that did not make contributions to candidates, parties, or other PACs could accept unlimited contributions from individuals, unions and corporations for the purpose of making independent expenditures."[23] As long as such PACs do not donate **directly** to a candidate, or coordinate their efforts **directly** with campaign staff, "there is no ceiling to how much money is injected into elections."[24]

Of course, Super PACs can, and do, spend massive funds "for such things as creating TV or radio ads supporting or excoriating particular candidates,"[25] or government policies. They just have to do it "independently." How independently is a bit cloudy, since "it is legal for candidates and Super PAC managers to discuss campaign strategy and tactics through the media." And "some Super PACs are run or advised by a candidate's former staff or associates."[26] The key word here being "former."

In the 2012 election, one single Super PAC spent $40 million to boost Republican candidate Mitt Romney's presidential bid.[27] Independently.

In the 2018 U.S. midterm elections, hedge fund owner Kenneth Griffin "gave $18.4 million to pro-Republican outside spending groups ... including $10 million to the New Republican PAC, a super PAC that spent $29 million on negative ads against former Democratic Sen. Bill Nelson (Fla.)."[28] Nor were Republicans the only beneficiaries. In the same 2018 midterms, "finance industry donors ... gave $264 million to Democrat-supporting outside groups."

Imagine the mafia being awarded "personhood," and thus all of its attempts to promote its activities, or use money to convince government that those activities should be legalized, being protected as "free speech." Imagine its dons, loan sharks and enforcers endowed with corporate immunity, never liable to individual punishment, with the only penalties for their bad behaviour being that a particular, local crime family had to pay a fine. They'd think they'd died and gone to heaven, or hell as the case may be.

The growth of elite, corporate control of governments outside the U.S. has followed similar pathways, yielding sometimes equal, sometimes lesser domination than in the U.S.

Canada, for example, though it too has PACs, has different federal and provincial campaign finance laws than the U.S., and as a result corporate domination there is slightly less blatant. As for the composition of its elites and their influence, three somewhat outdated but still excellent examinations of the Canadian situation are: *The Insiders: government, business and the lobbyists*, by John Sawatsky,[29] Peter C. Newman's two-volume *The Canadian Establishment*,[30] and *The Canadian corporate elite: an analysis of economic power*,[31] by Wallace Clement.

Whatever the exact degree of national, state, provincial or local control, however, corporate rule is the general rule across most of today's world. And in most of the world it is leading to disaster. The problem is stopping it, and stopping it in time. Much more delay and the planet could be too far past several key physical tipping points, and already in its death throes.

What cannot be done

To try to outspend the millionaire class, their multibillion-dollar corporations, their lobbyists, public relations specialists, and their proliferating Super PACs in buying political influence is an obvious mugs' game which ordinary people with limited financial means stand no chance of winning. As for mass organizing of voters via traditional political parties, with the goal of bringing enough counter-pressure to change government policy, it may gain a few, minor improvements. But it takes years of effort. To bring about the major changes required to rescue the planet, by going through the standard electoral process, would take many decades.

Decades the environment, and the human species, no longer have.

Nor is much success likely to come from simply pointing out the damage a corporation is doing, and attempting via "moral suasion" to convince its directors to change their policies. This is because, as a growing number of observers have noted, the peculiar sort of greed displayed by some of the corporate elite, and their indifference to the destructive results of their actions, is pathological. They are too often not well-balanced people, with whom one could expect to reason.

For more than two decades, evidence has been piling up that there is such a thing as "executive psychopathy," and that many of those so afflicted are actually suffering from a form of addiction—not to heroin or crack cocaine, but to money and power, and the high that comes with them. One such addict, now recovering, described in the *New York Times*[32] what it is like:

> *In my last year on Wall Street my bonus was $3.6 million—and I was angry because it wasn't big enough. I was 30 years old, had no children to raise, no debts to pay, no philanthropic goal in mind. I wanted more money for exactly the same reason an alcoholic needs another drink: I was addicted ... I was a giant fireball of greed ... I wanted a billion dollars. It's staggering to think that in the course of five years, I'd gone from being thrilled at my first bonus—$40,000—to being disappointed when, my second year at the hedge fund, I was paid "only" $1.5 million.*
>
> *The author, a derivatives trader, added: "I noticed the vitriol that traders directed at the government for limiting bonuses after the [2008] crash. I heard the fury in their voices at the mention of higher taxes. These traders despised anything or anyone that threatened their bonuses. Ever see what a drug addict is like when he's used up his junk? He'll do anything—walk 20 miles in the snow, rob a grandma—to get a fix. Wall Street was like that. In the months before bonuses were handed out, the trading floor started to feel like a neighbourhood in* The Wire *when the heroin runs out."*

Not only Wall Street, but much of the corporate world, is dominated by such people. Hedge fund investors, for example, now control many non-financial businesses, and those hedge funds may in turn be dominated by the sort of "greed addicts" the author, and the journalist who wrote of "rich white families" above, describe.

The phenomenon comes under many titles, including "corporate psychopathy," "executive psychopathy," "antisocial personality disorder," "sociopathy," "narcissistic personality disorder," "addictive greed/power dysfunction," and so on. The various descriptive titles stem from sometimes minor, sometimes major variations in behavior, as well as in the antecedents that gave rise to each condition.

For example, psychopathy in general is described by the *Wikipedia* as "a personality disorder characterized by persistent antisocial behaviour, impaired empathy and remorse, and bold, disinhibited, and egotistical traits."[33] Noting it is "sometimes considered synonymous with sociopathy," the description adds that "different conceptions of psychopathy have been used throughout history that are only partly overlapping and may sometimes be contradictory ...

"It may be that a significant portion of people with the disorder are socially successful and tend to express their antisocial behavior through more covert avenues such as social manipulation or white collar crime. Such individuals are sometimes referred to as 'successful psychopaths,' and may not necessarily always have extensive histories of traditional antisocial behavior as characteristic of traditional psychopathy."[34]

The *Psychology Today* website[35] is more specific:

"Psychopathy is among the most difficult disorders to spot. The psychopath can appear normal, even charming. Underneath, he lacks conscience and empathy, making him manipulative, volatile and often (but by no means always) criminal. She is an object of popular fascination and clinical anguish; adult psychopathy is largely impervious to treatment, though programs are in place to treat callous, unemotional youth in hopes of preventing them from maturing into psychopaths.

"Psychopathy is a spectrum disorder [one 'that includes a range of linked conditions ... symptoms and traits'[36]] and can be diagnosed only using the 20-item Hare Psychopathy Checklist. (The bar for clinical psychopathy is a score of 30 or more). Brain anatomy, genetics and a person's environment may all contribute to the development of psychopathic traits.

"The terms 'psychopath' and 'sociopath' are often used interchangeably, but in correct parlance a 'sociopath' refers to a person with antisocial tendencies that are ascribed to social or environmental factors, whereas psychopathic traits are more innate, though a chaotic or violent upbringing may tip the scales for those already predisposed to behave psychopathically. Both constructs are most closely represented in the *Diagnostic and Statistical Manual of Mental Disorders (DSM)* as Antisocial Personality Disorder."

Writing in the *Psychology Today* blog, Dr. William Hirstein provides some historical background for the terms:

"In the early 1800s, doctors who worked with mental patients began to notice that some of their patients who appeared outwardly normal had what they termed a 'moral depravity' or 'moral insanity,' in that they seemed to possess no sense of ethics or of the rights of other people. The term 'psychopath' was first applied to these people around 1900. The term was changed to 'sociopath' in the 1930s to emphasize the damage they do to society. Currently, researchers have returned to using the term 'psychopath.' Some of them use that term to refer to a more serious disorder, linked to genetic traits, producing more dangerous individuals, while continuing to use 'sociopath' to refer to less dangerous people who are seen as products of their environment."[37]

Two influential business magazines suggest why such people often turn up in corporate executive suites:

Harvard Business Review: "Many of psychopaths' defining characteristics—their polish, charm, cool decisiveness, and fondness for the fast lane—are easily, and often, mistaken for leadership qualities. That's why they may be singled out for promotion. But along with their charisma come the traits that make psychopaths so destructive: they're cunning, manipulative, untrustworthy, unethical, parasitic and utterly remorseless. There's nothing they won't do, and no one they won't exploit, to get what they want. A psychopathic manager with his eye on a colleague's job, for instance, will doctor financial results, plant rumors, turn coworkers against each other, and shift his persona as needed to destroy his target. He'll do it, and his bosses will never know."[38]

Forbes magazine: "Psychopaths may be charismatic, charming, and adept at manipulating one-on-one interactions. In a corporation, one's ability to advance is determined in large measure by a person's ability to favorably impress his or her direct manager. Unfortunately, certain of these psychopathic qualities—in particular charm, charisma, grandiosity (which can be mistaken for vision or confidence) and the ability to 'perform' convincingly in one-on-one settings—are also qualities that can help one get ahead in the business world."[39]

Combine these traits with those associated with addiction and you have the potential for disaster. Addiction is described as "a condition in which a person engages in use of a substance *or in a behavior* for which the rewarding effects provide a compelling incentive to repeatedly pursue the behavior despite detrimental consequences ... the addictive substances and behaviours share a key neurobiological feature—they intensely activate brain pathways of reward and reinforcement, many of which involve the neurotransmitter dopamine."[40] [More on addiction, particularly addiction to violence, in the following chapter.]

There certainly isn't much potential for practical, rational dialogue with such people, any more than earlier generations had for rational dialogue with Hitler and his key elite. The Nazis were, as Norman Ohler shows in his 2016 book *Blitzed: drugs in Nazi Germany*,[41] not only power addicts but frequently hard drug addicts as well—Hitler on methamphetamine and cocaine, and Goering on heroin.

There is thus far no evidence that the current U.S. president is addicted to drugs, but his wilder inclinations may indicate an unhealthy attraction to power that has some in his entourage worried. A recent *New York Times* article, purported to be by one of this inner circle, claims that his own appointees work surreptitiously to "frustrate parts of his agenda and his worst inclinations," which include "half-baked, ill-informed and occasionally reckless decisions."[42] Even at the political top, apparently, subterfuge rather than reason prevails.

~

If you can't outspend today's corporate elites in buying normal political influence, or appeal to their consciences by reasoning with them, what *can* you do?

It appears there are very few options:

1) bring such overwhelming pressure on governments that, despite the elites' entrenched financial influence, they will compel the corporate world to reform and reverse course. Unfortunately, as already noted, in today's emergency context this could take far too much time, or require so much force as to constitute a virtual overthrow of government;
2) bring sufficient direct pressure on the corporate CEOs and directors themselves to force them to change course without government coercion , or,
3) simply remove the worst corporate offenders from their posts and replace them with people who will act responsibly. In the latter case, this may mean some form of public and/or worker control of corporate boards, or that bogeyman of every capitalist —nationalization.

Finding ways to do what must be done, including whether to use violent or nonviolent methods, are the subjects of the next chapter.

---

1 Norman Ohler, *Blitzed: drugs in Nazi Germany*, (Allen Lane, an imprint of Penguin Random House UK, 2016), 276.
2 Judy Pearsall and Bill Trumble, eds., *The Oxford English Reference Dictionary* (Oxford: Oxford University Press, 1996, 338.
3 Farlex, "Organized crime: legal definition of organized crime," *The Free Dictionary*, as posted online 8 August 2018 at https://legal-dictionary.thefreedictionary.com/Organized+Crime.
4 John H. Cushman Jr., "Shell knew fossil fuels created climate change risks back in 1980s, internal documents show," 5 April 2018, *Inside Climate News*, as posted online 5 April 2018 at https://insideclimatenews.org/news/05042018/shell-knew-scientists-c ...
5 Carey Gillam, "One man's suffering exposed Monsanto's secrets to the world," 11 August 2018, *The Guardian*, as posted online 11 August 2018 at www.theguardian.com/business/2018/aug/11/one-mans-sufferi ..., 2.
6 *Op. Cit.*, 1.
7 Desert Hope Treatment Center, """Big pharma," as posted online August 2018 at https://deserthopetreatment.com/big-pharma/, 7.
8 *Loc. Cit.*
9 *Op. Cit.*, 8.
10 Jessica Glenza, "US opioids: number of addicted women giving birth quadrupled over 15 years," -numbe ...
11 Andrew Joseph, "Richard Sackler, member of family behind OxyContin, was granted patent for addiction treatment," 7 September 2018, *STAT News*, as posted online

7 September 2018 at www.statnews.com/2018/09/07/richard-sackler-member-of-family-behind-oxycontin-was-granted-patent-for-addiction-treatment/

12 Chris McGreal, "Billionaire founder of opioid firm guilty of bribing doctors to prescribe drug," 2 May 2019, *The Guardian*, as posted 2 May 2019 at www.theguardian.com/us-news/2019/may/02/john-kapoor-o ..., 1.

13 *Loc. Cit.*

14 Allan J. Lichtman, "Who rules America?" 8 December 2014, *The Hill*, as posted online 6 August 2018 at http://thehill.com/blogs/pundits-blog/civil-rights/214857-who-rules-america, 1.

15 Richard Harwood, "Monopoly by a single class undermines the system," November 1994, *The Washington Post*, reprinted by the *New York Times* on its Op-Ed page.

16 16 *Wikipedia, the Free Encyclopedia*, "Hedge fund," as posted online 14 August 2018 at https://en.wikipedia.org/wiki/Hedge_fund, 1.

17 G. William Domhoff, "Who rules America?" *WhoRulesAmerica.net*, as posted online 5 August 2018 at https://whorulesamerica.ucsc.edu/power/class_domination.html?print, 4.

18 Chris Hedges, "The pathology of the rich white family," 18 May 2015, *Truthout*, as posted online 12 August 2018 at www.truthdig.com/articles/the-pathology-of-the-rich-white-fam ..., 1.

19 *Op. Cit.*, Domhoff, "Who rules America?"

20 *Wikipedia, the Free Encyclopedia*, "Corporate personhood," as posted online 2 August 2018 at https://en.wikipedia.org/wiki/Corporate_personhood, 1.

21 *Op. Cit.*, 3.

22 *Loc. Cit.*

23 *Wikipedia, the Free Encyclopedia*, "Political action committee," as posted online 12 August 2018 at https://en.wikipedia.org/wiki/Political_action_committee #Super_PACs, 2.

24 Chris Warren, "How Super PACs work," *howstuff works.com*, as posted online 12 August 2018 at https://howstuffworks.com/super-pac1.htm, 2.

25 *Loc. Cit.*

26 *Op. Cit.*, *Wikipedia*, "political action committee," 2.

27 *Loc. Cit.*

28 Donald Shaw, "Hedge fund billionaires were Democrats' main bankrollers in 2018," 17 May 2019, *Truthout*, as posted online 17 May 2019 at https://truthout.org/articles/hedge-fund-billionaires-were-democrats-main-bankrollers-in-2018/, 1.

29 John Sawatsky, *The insiders: government, business and the lobbyists*, (Toronto: McClelland and Stewart, 1987).

30 Peter C. Newman, *The Canadian establishment*, (Toronto: McClelland and Stewart, 1975 Vol. 1, 1981 Vol. 2)

31 Wallace Clement, *The Canadian corporate elite: an analysis of economic power*, (Ottawa: Carleton University Press, 1986)

32 Sam Polk, "For the love of money," 18 January 2014, *The New York Times*, as posted online 13 August 2018 at www.nytimes.com/2014/01/19/opinion/sunday/for-the-love-of-money.html, 1-2.

33 *Wikipedia, the Free Encyclopedia*, "Psychopathy," as posted online 21 April 2018 at https://en.wikipedia.org/wiki/Psychopathy, 1.

34 *Op. Cit.*, 4.

35 Psychopathy Basics, "What is psychopathy?" *Psychology Today*, as posted online 21 April 2018 at www.psychologytoday.com/us/basics/psychopathy

36 *Wikipedia, the Free Encyclopedia*, "Spectrum disorder," as posted online 20 August 2018 at https://en.wikipedia.org/wiki/Spectrum_disorder, 1.

37 William Hirstein, "What is a psychopath?" 30 June 2013, *Psychology Today*, as posted online 12 May 2017 at www.psychologytoday.com/blog/mindmelding/201301/what-is ..., 1.

38 Gardiner Morse, "Executive psychopaths," October 2004, *Harvard Business Review*, as posted online 21 April 2018 at https://hbr.org/2004/10/executive-psychopaths, 1.
39 Victor Lipman, "The disturbing link between psychopathy and leadership," 25 April 2013, *Forbes*, as posted online 21 April 2018 at www.forbes.com/sites/victorlipman/2013/04/25/the-disturbing- ..., 1.
40 *Psychology Today*, "What is addiction?" as posted online 20 August 2018 at www.psychologytoday.com/us/basics/addiction
41 Ohler, *Op. Cit., Blitzed*
42 Ben Jacobs, "Trump cries 'treason' as senior official attacks president in anonymous NYT op-ed," 6 September 2018, *The Guardian*, as posted online 6 September 2018 at www.theguardian.com/us-news/2018/sep/05/donald-trump- ...

Chapter Seven

WHAT WORKS?

"Violence and hate only breed violence and hate ... violence is impractical."
—Martin Luther King Jr., *Autobiography*[1]

"Aikido ... is the protector of all living things."
—Ueshiba Morihei, founder of Aikido[2]

"It's an important job, as long as one cares about the end, and not too much about the means."
—George Smiley, in John LeCarre's *A Legacy of Spies*[3]

A disclaimer ...

From what the previous chapters have shown, our planet is now facing the worst crisis in human history, one that threatens not only the future of our own species, but that of most other currently living things. Not since the Great Dying of the dinosaurs 65 million years ago has anything like it been seen.

Given the extreme situation, it would be foolishly presumptuous for any author to pretend to know what is the "right" thing for each of us to do. Every individual must consult his or her own conscience—and consult it more seriously than ever before—then act accordingly.

The only position I would argue absolutely against is not to act, to do nothing. Inaction, when everything we could possibly hold dear is threatened—including our own children, if we have any, and all other children, all birds, animals, trees, plants, fish, insects—would, I believe, be truly sinful, if such a concept has any meaning at all. It would be to judge not only ourselves, but all things to be without value, not worth the bother to defend.

It would also be to decline to take part in the greatest adventure, the greatest cause, ever on offer.

To do so, knowing that a small minority of our own species, a cabal of mental and moral cripples, is responsible and might yet be restrained, would make us accomplices, as guilty as they are. They, at least, may not be in their right minds, addicts in the grip of their fixations, to be pitied. Most of us don't have that excuse.

But neither do most of us have the previous individual experience to afford certainty that we are on a right path. Here, the example and the words of those who have gone before us can be of help, especially those who faced dire crises of their own, from war and famine, to slavery, to sometimes brutal labour strife. This chapter will lean heavily on their thinking.

First, however, recall the fundamental goal: ***to stop the major corporations from continuing to destroy the key systems on which our lives depend and, once they have been stopped, to force them to help repair or reverse the damage they've already caused. That means targeting those who control the corporations, their CEOs, directors, and stockholders.***

> **As noted in the previous chapter, targeting governments—which in many countries are little more than the enforcement arm of the corporations, whose money buys their politicians—may have some slow, minimal effect, but seems more like trying to stop a tank by arguing with its gun barrel, rather than with the soldiers driving the machine.**

The corporate structure itself is a basic problem, but the worst individual corporate offenders, primarily responsible for the danger, are the chief pharmaceutical, chemical, agricultural, financial and energy (particularly fossil fuel) companies. The leading firms in each field are listed every year in such places as *Standard and Poor's* directory, *Fortune Magazine's* Fortune 500 list, and other such reference sources, while the names of their CEOs and boards of directors are published in the individual companies' annual reports. Stock brokerages and exchanges can indicate the major investors in each company. Most of this information is also available online.

Once these organizations have been brought under responsible (to the environment and to life) control, consideration can be given to ways of reversing their destruction, and—eventually—building the kind of society that will make such horrors relics of the past. There is an abundant literature on the sorts of social organization that might fill the latter bill, from the direct democracy favoured by theorists like Noam Chomsky or Murray Bookchin (see Chapter 8) to the dream of the International Workers of the World (IWW) of "an industrial commonwealth in which all governmental functions as we know them today will have ceased to exist, and in which each industry will be controlled by the workers in it without external interference."[4] (See below for more on the IWW).

That, however, must be the subject of another book, not this one. We are all, right now, under violent, potentially lethal attack by our onrushing corporate aggressors. The immediate task, the job at hand, in words familiar to every *aikidoka* (practitioner of the martial art of Aikido, described below), is: Step in, turn.

A just war

Obviously, young people—the so-called "millennials" and those born after them—should take the lead in stepping in, since they are the ones with the most to lose. People this author's age (78 at this writing) will likely already be dead, or ga-ga with dementia, before the worst happens, and as Chapter Three illustrated, the young have already begun to suffer, via the unfair student loan system and the debt it has saddled them with. How did their predecessors handle things? Whose example might show them the way, or ways?

Of crucial importance is the question of violence versus nonviolence.

A "shooting" war, of course, is the ultimate in human violence, while the coming battle could surely be seen, in many ways, as a moral or social "war." This necessarily brings up the concept, studied and argued over for centuries by philosophers and theologians, of the "Just War" (*jus bellum*).

The Roman senator and philosopher Marcus Tullius Cicero (106-43 BC) was among the earliest, and best known, thinkers to deal with the question, notably in his book, *On Obligations* (Latin *De officiis*), addressed as advice to his son, Marcus. His words on almost any subject obviously carried weight with his contemporaries, to the point where his eventual assassins not only cut off his head, but also his hands (for writing the truth).[5] His thinking on justice dovetailed closely with his thoughts on war, and actually paved the way for them.

"There are two kinds [of injustice]:" he wrote, "the injustice of those who inflict it, and that done by those who do not protect victims from injury when they have the power to do so ... the man who does not repel or oppose some wrong when he can do so, is as much at fault as if he were abandoning parents, friends or country."[6]

He believed such "protection" could include war. "At the level of state policy ... wars should be undertaken for the one purpose of living peaceably without suffering injustice; and once victory is won, those who have not indulged in cruel monstrosities in the war should be spared." As for treaties following wars, he added: "We should always aim at a peace which does not contain the seeds of future treachery."[7]

As general advice, it sounded good, but left enough holes to drive several chariots through. What, for example, would constitute an "injustice" at "state level?" How, exactly, might one define a "cruel monstrosity?" Later thinkers attempted to fill in such blanks, from the Christian saints Augustine and Thomas Aquinas, to the so-called School of Salamanca, which added the crucial caveat that war "ought to be resorted to only when it was necessary in order to prevent an even greater evil."[8]

The school's theorists also required that "governing authorities" must declare war, but "if the people oppose a war, then it is illegitimate. The people have a right to depose a government that is waging, or is about to wage, an unjust war." Precisely how many individuals would constitute "the people," and how their consent might be obtained was not detailed.

They also added that "one may not attack innocents or kill hostages." Would the current American government's stance on "collateral damage" (namely, that killing civilians is excusable as long as they are not the main object of an attack) violate this rule? And would not atomic war, and the resulting destruction of most of the planet, be a "greater evil" than any it could prevent?

The best criteria of "just cause" for war that this author has found were outlined in the old reliable *Wikipedia*, in its entry on "Just War Theory."[9] But any theory—even the most detailed and carefully-worded—can still fall apart on the details.

For example, at the time it began, the First World War, which killed at least 17 million people and wounded some 20 million more, while devastating most of Western Europe, was described as "just" by any number of its apologists on both sides, in and out of government, even though, today, a clear actual reason for launching that slaughter has yet to be agreed upon by historians. "The causes of World War I remain controversial," says the *Wikipedia*, in obvious understatement. The war also maimed or killed millions of animals, from the horses who pulled cannons and the "mercy dogs" who found wounded soldiers, to the pigeons who carried messages and dispatches. An estimated 484,143 horses, mules, camels and bullocks were killed in British service alone between 1914 and 1918,[10] and no one, then or now, really knows why.

Nor are the people and animals killed or wounded during a war, and the property destroyed, its ultimate measure. The long term, knock-on effects, including postwar social and psychological ones, should not be overlooked. The American Civil War, for example, dwarfed even the Napoleonic Wars in sheer destructiveness, outmatching anything in the 19th century. But its evils didn't end in 1865, as the makers of the 1993 film, *The Real Story of Butch Cassidy and the Sundance Kid*,[11] explained.

"Although the war was over, its legacy of killing was to have a lasting impact on the attitudes and convictions of millions of Americans," says the film's narrator, to which Emory University historian Michael A. Bellesiles adds:

"Any legal historian who has studied the court records or criminal records of the *anti-bellum* United States knows that murder was not a common crime. As a matter of fact, it was quite exceptional, and any murder would be a front-page story in whatever community it appeared in. All this changes with the Civil War. While white settlers didn't hesitate to kill African-Americans or native Americans, what the Civil War does is two things.

"First, it teaches white Americans that it's quite acceptable to murder other white Americans. We have to remember the Civil War was four years of unrelenting bloodshed, carnage, murder on a mass scale. No one in the 19th century ever had seen anything like it. The second major change effected by the Civil War was that it armed America. Congress in 1865 passed an act allowing all union soldiers to keep their personal weapons, to take them home.

"In terms of crime, the Civil War is the great cultural divide in the United States. Prior to the Civil War, crimes in the U.S. were committed much as they were in England, and are to this day, without weapons. After the Civil War, crime [in the U.S.] became a matter of violence. At the very least, a criminal pretty much had to carry a gun because, after all, he didn't know who else was armed."

The film narrator adds that while many desperados hesitated to kill, for fear of being hanged later, "the lawmen, especially those who had served in the Civil War, just did not have this hesitance. They were quite willing to open fire, almost at random. After all, that had been their experience in war. It had been encouraged, and they had even been rewarded for these actions."

The result was the "Wild West" of song and story.

Historically, few wars or violent revolutions have succeeded in replacing whatever social or political evil they aimed to end with anything socially or politically better, at least in the short run. The violence in France in 1789 toppled the decadent *Ancien Régime*, but ended with the dictatorship of Napoleon and the Napoleonic Wars, up till then the most destructive wars Europe had seen. The violence of the Bolsheviks brought down the despotic Russian Tsars, but brought in the dictatorship of Stalin, the oppression of Eastern Europe, the Cold War and the "balance of terror," which came close at least once, during the Cuban missile crisis, to launching nuclear oblivion.

Addiction, again

The trouble is, violence on any scale, whether the large one of international warfare or the small one of group or individual struggles, has an ancillary, long-term mental impact. Much like greed or power lust, discussed previously, it is addictive. People have known for a long time, on a common-sense level, that the victims of violence aren't the only ones affected, whether physically or via the psychological effects of Post Traumatic Stress Disorder (PTSD).

The perpetrators can also be changed, and not necessarily for the better. They come to like it.

Why this should be still awaits a definitive, detailed scientific explanation. But scientists are on the trail of the physical, neurological causes, and many of the mechanisms involved are beginning to be understood. Some of the most recent research publications on the subject are listed in the suggested readings in Chapter Eight of this book.

Briefly, the human brain reacts not only to ingested substances like food or drugs, but also to physical and psychological experiences, which can trigger various brain centers and the circuits that connect them. Biochemicals that flow along those circuits may stimulate or damp down activity in the centers they encounter en route.

Circuits in both the amygdala and hypothalamus may promote aggression if stimulated, while the frontal cortex tends to mediate it. "The frontal cortex provides inhibitory inputs to circuits in the hypothalamus and amygdala that might promote aggression ... many studies have reported a link between brain damage to the frontal cortex and increased aggressive behaviour."[12]

Oxytocin, a neuropeptide hormone, can reduce activity in the amygdala, producing a calming effect. Dopamine, referred to as the "pleasure chemical" or "reward chemical" in many popular press articles, may have numerous effects. Among them, is a tendency to reward some negative activities, like the greedy accumulation of money or power, or the commission of violent acts, with a "high." Testosterone can also play a part, acting on even those who merely observe conflict or aggression, rather than on those directly involved in it as victim or perpetrator:

"Sports fans respond to watching their team win or lose with corresponding increases or decreases in testosterone levels. Children playing violent video games show reduced activation of brain areas involved in affect [ability to feel emotion, including sympathy towards others], such as the amygdala and the anterior cingulate cortex. Reduced brain activity in frontal areas has also been reported in children with high exposure to violent video games and television programs."[13]

The whole process of producing dopamine, testosterone or other substances, and their often contradictory influences in various brain centers, is extremely complicated, and should not be oversimplified. Suffice it to say, however, that violence is a very risky business, psychologically as well as physically, with frequently unexpected side effects on everyone involved.

The perpetrators may turn into "war lovers," like the serial-killer character played by John Saxon in the film, *War Hunt*,[14] or thrill seekers, or just plain bullies, worse than any of the sociopaths they may defeat. And of course, they may provoke a burning desire for vengeance in their victims, including those harmed as part of what has euphemistically come to be called "collateral damage." Much of the anti-Western hatred in Middle Eastern countries has stemmed from such damage.

Violent revolts, guillotines, firing squads, murders in locked rooms, may satisfy impatience or the desire for vengeance, but they establish a habit of mind, and tend to make leaders of despots. And of course they leave a core of angry, embittered people, dreaming of getting their own back: the losing side, and their relatives or friends.

If employed at all, violence should be the absolute last resort, and the utmost care taken to prevent injury to the innocent, whether human or animal. To deliberately risk the injury or death of children, as "disgruntled," underpaid seasonal workers on Australian strawberry farms did when they put steel needles in the fruit they picked,[15] is inexcusable. A "7-year-old girl in South Australia state found a needle in a Western Australia-grown strawberry."[16]

Some foreign workers had been underpaid and at least one grower was "convicted and fined almost $70,000 for underpaying a number of his staff at his former strawberry farm in Stanhope,"[17] Australia. The workers often put in "16-hour days with minimal breaks." Those workers may have had a legitimate grievance, but chose a totally illegitimate way to express it. The pain or death of a child, or anyone, would wipe out any justification. And, of course, would provoke resentment and anger in observers who otherwise might have been sympathetic to the fruit pickers' cause.

Even relatively minor, but negative, impacts on innocent or uninvolved victims of protest can be counter-productive. For example, the now well-known group calling themselves Extinction Rebellion (ER) launched a series of "swarmings" to block public streets in London, England, to call attention to the "threats of climate change and species extinction." They walked into the streets at key intersections during morning rush hour, causing major traffic jams and delaying people trying to get to their workplaces.[18]

They unquestionably drew attention, but it might have been counter-productive, if the attitudes of some motorists were indicative. "Cars sounded their horns, and one driver got out to remonstrate with police monitoring the protesters. Officers told him to be patient. Another passer-by, who asked police why they could not clear the roadblock, was told the activists had a right to protest. 'You're part of the problem,' he shouted back at them."[19] Another motorist said he "couldn't give a damn" about their issue.

The corporations actually responsible for climate change and species extinction were likely pleased, to see their opponents giving themselves a bad name with the public, and to know that the protesters' goal, as expressed in a public letter, was to "support rebellion against the UK government," rather than against the corporations.

The billionaire CEOs and millionaire directors could assume that, if such protests should grow unruly enough, they may spook the politicians into imposing something like a carbon tax on ordinary, rank-and-file citizens (who would then have even more reason to resent the protesters), but not likely into biting the corporate hand that feeds their election fund coffers, and provides for their comfortable retirement when they leave office.

Vandalism and property damage are also forms of violence, potentially as addictive as physical violence against people or other living things. And, just as there are criminal and civil laws against harming people or animals, there are laws against damaging property. These vary from jurisdiction to jurisdiction, but criminal penalties in the U.S. state of Michigan are fairly typical. The *Justicia* blog posted by the law firm Hilf & Hilf, PLC, gives an excellent outline.[20]

At the low end of the scale, doing $200 or less damage is rated as a misdemeanour and can call for a penalty of 93 days in jail and a fine of $500. At the high end, damage of more than $20,000 is a felony and can net a 10-year prison sentence and fine of "$10,000 or three times the amount of the destruction, whichever is greater."[21]

And, of course, though the blog doesn't mention it, the perpetrators damage themselves psychologically. "Firebugs," who can't stop themselves from setting fires, are as much addicts as anyone taking heroin, and like every addict, started with a single "dose."

"Thoughtcrime" and punishment

As important as the question of whether or not to use violence is whether or not to endure it, or to endure imprisonment, as a result of dissenting. The measure of one's exposure to such dangers depends on where one chooses to act. To paraphrase George Orwell in *Animal Farm*, some venues are "more equal than others" in the extent to which they belabour protestors. And some are also far more equal in what they consider punishable offences. In some countries, particularly the U.S. (as will be shown below), the most recent laws verge on Big Brother's rules against "thoughtcrime" in Orwell's other classic, *Nineteen Eighty-Four*.[22]

"Once upon a time," the U.S., Canada, England and France were among the world's bastions of political freedom of expression, as well as of direct action to secure justice and "the redress of grievances."[23] Not only did their basic documents, from Britain's *Magna Carta* (1215) and the U.S. Constitution (1789) to *France's Declaration of the Rights of Man* (1789), guarantee basic freedoms, but long legal practice, built up over decades and even centuries, had expanded them.

The latter often came as a result of fierce social battles, such as the 19th century labour strikes and "*sabotage*" (working slowly and clumsily, as if wearing wooden shoes, or *sabots*) in France, where the

Congrès confédéral de Toulouse first officially endorsed such tactics, in 1897,[24] or the wild violence of the early mine workers' strikes in Kentucky, USA, and the famed Battle of the Overpass in Dearborn, Michigan, where the fledgling United Auto Workers (UAW) union was attacked by strike-breaking thugs.

As a young university student, this author heard the tale of the miners' fight from the grandson of one of the strikers. The grandson, who hailed from the legendary Harlan County, and I sat up late at night in the university dormitory, as he repeated the tale, handed down in his family, of how strike-breakers came with mounted machine guns and fired at the miners, who refused to flee. Later, as a reporter for a Detroit newspaper, I heard the story of the UAW's Overpass fight directly from UAW founder and president Walter Reuther. So entranced was I by Reuther's narrative at the time that I forgot to take notes, but Reuther's recollections were reported later by another journalist:

UAW workers had begun passing out leaflets at a pedestrian walkway near the Ford Rouge plant in Dearborn, Michigan, in May 1937, when hired strike-breakers attacked them and began beating them with fists and clubs. Reuther (speaking to a group of reporters in 1997) recalled: "Seven times they raised me off the concrete and slammed me down on it. They pinned my arms, and I was punched and kicked and dragged by my feet to the stairway, thrown down the first flight of steps, picked up, slammed down on the platform and kicked down the second flight. On the ground they beat and kicked me some more ... "[25]

Reuther survived without serious injury, although another union organizer suffered a broken back in the incident. Ford Motor Company, however, paid for its sins, receiving a rash of bad national publicity as a result of the violence, and eventually was forced to sign a contract with the union.

Other freedoms, as well as innovative direct-action tactics used to win them, came from other battles, including the struggles of the French *Resistance* against Nazi occupation in the Second World War. There, *Maquis* leader Jean Moulin[26] and his fellow fighters used an array of methods. Typical was the decision—when the German invaders began to force French workers to ship out to war production plants in

Germany—to *volunteer* for transport to the plants. Resistance fighters made the trip, and once on site began to organize extensive acts of sabotage to foul up German war production. Some were later caught and shot dead. Moulin himself wasn't among the infiltrators, but was eventually captured and tortured to death by the infamous *Gestapo* officer Klaus Barbie.

The rights won as a result of so much sacrifice were substantial, but not without some fairly clear limits. The key phrase in the U.S. Constitution's First Amendment, for instance, was "*peaceably* to assemble," assemblies of any other kind being deemed illegal, possibly constituting the crime of rioting.

Common definitions of the word "riot" include *Webster's College Dictionary's*[27] "a noisy, violent public disorder caused by a group or crowd of persons," the *Oxford English Reference Dictionary's* "a disturbance of the peace by a crowd,"[28] and the online *Free Dictionary's* more specific "a disturbance of the peace by several persons, assembled and acting with a common intent in executing a lawful or unlawful enterprise in a violent and turbulent manner."[29]

Under U.S. federal law, a "riot" was until very recently defined as: "A public disturbance involving (1) an act or acts of violence by *one or more* persons [thus a one-man riot was theoretically possible], which act or acts shall constitute a clear and present danger of, or shall result in, damage or injury to the property of any other person or to the person of any other individual or (2) a threat or threats of the commission of an act or acts of violence by one or more persons part of an assemblage of three or more persons having, individually or collectively, the ability of immediate execution of such threat or threats, where the performance of the threatened act or acts of violence would constitute a clear and present danger of, or would result in, damage or injury to the property of any other person or to the person of any other individual."[30]

M. Cherif Bassiouni outlined U.S. requirements as they stood in 1971, along with their historical development, in his excellent, detailed study *The Law of Dissent and Riots.*[31] Noting that a variety of federal, state and even local laws put boundaries of their own (in addition to federal law) on what was permitted, he explained that what legally constituted a riot, as well as the punishment for participating in one, varied widely from one U.S. state to another.

For example, a group of two or three persons who commit or threaten to commit violence can legally constitute a riot in most states. As of 1969, three people were required in 31 states, two in 10 states. But some loosey-goosey locales like Pennsylvania or Rhode Island (12 persons) or Kentucky (20) were far more permissive. Michigan required no fewer than 30 unarmed participants, but only 12 armed ones![32] As for punishments, these ranged from a low of 30 days in jail plus a $500 fine in Ohio or South Carolina, to two years in jail and a fine of $2,000 in Arizona and Montana.[33] A welter of local municipal laws might also come into play, from prohibitions against blocking traffic to limits on noise, such as loud music or voices shouting through bullhorns.

In other countries, the laws and definitions may be more or less specific, and more or less strict. In England and Wales, the *Wikipedia* notes that the Public Order Act states "where 12 or more persons who are present together use or threaten unlawful violence for a common purpose and the conduct of them (taken together) is such as would cause a person of reasonable firmness present at the scene to fear for his personal safety, each of the persons using unlawful violence for the common purpose is guilty of riot."[34] A person convicted of such "is liable to imprisonment for any term not exceeding 10 years, or to a fine, or to both."

In short, anyone planning a public demonstration had better check the statute books beforehand, or better yet a lawyer, to see if what they plan could land them in jail.

In the United States today, they had better check the most recent statutes, because since the terrorist attacks in New York in 2001 the laws protecting not only free speech and assembly, but any and all human rights have been relentlessly weakened. Under former President Bush, and now under Trump, they continue to be weakened further, to the point where applying the term "law" to them seems almost an exaggeration.

Under the guise of fighting "terrorism," two federal attorneys general have issued guidelines limiting the extent to which agencies like the U.S. Federal Bureau of Investigation (FBI) can spy on, or eventually charge political dissenters with crimes. Various other branches of government have also loosened control over treatment of peaceful dissenters.

Shortly after the 9/11 terror attacks in New York, then-Attorney General John Ashcroft issued guidelines "that provided for wholesale political spying on dissenters" including previous rules "that had barred the FBI from attending political meetings and houses of worship to spy on activities and individuals not suspected nor accused of any crimes. The Ashcroft guidelines instituted massive surveillance of political meetings and rallies, religious gatherings, Internet sites and bulletin boards, and other purely expressive activities explicitly protected by the First Amendment. The Ashcroft guidelines also authorized the FBI to enter massive numbers of names of individuals under such surveillance into government databases."[35]

As bad as this was, Ashcroft's successor Michael Mukasey in 2008 went further, issuing a new set of guidelines[36] that "do not limit the FBI but afford it such untrammelled power that the guidelines might as well not exist. Under the Mukasey guidelines, the FBI may investigate anyone at all, even in the absence of any evidence whatsoever of a crime. The FBI is authorized to investigate political demonstrations, is newly permitted to employ a variety of intrusive investigative techniques previously off-limits, and is actually encouraged to make aggressive use of informants ...

"The bureau now has the authority to investigate 'threats to the national security,' an ill-defined term that has been interpreted to cover all forms of political dissent, making the FBI into a kind of political police ... [Anyone with any connection to a foreign policy issue can now legally be investigated], even a professor writing about a foreign country ... anyone who voices an opinion on almost any world issue is subject to investigation, recruitment and record keeping."[37]

If any of this near-paranoid surveillance results in an arrest and/or criminal charge, the dissenter should be worried, as "a federal felony [is] punishable with five to eight years in prison."[38]

What might constitute a chargeable offense has also been changed since 9/11, under the Bush administration's Patriot Act: "Employing an absurdly broad definition of 'domestic terrorism,' the [act] turns almost all forms of vigorous protest and minor criminal conduct into prosecutable acts of terrorism."[39]

And today's Trump administration has demonstrated, via multiple public statements and actions, such disregard for truth or law that ordinary citizens might be tempted to fear arbitrary arrest or charges for nearly any activity. The cry "lock her up," originally directed at presidential candidate Hillary Clinton, could as easily be directed at anyone. Journalists, branded "enemies of the people" by Trump, would seem particularly vulnerable, especially those covering foreign affairs, who under the new rules could at least theoretically be deemed "terrorists" simply for writing about any foreign country.

They could also be murdered, if they should turn up on the U.S. government's secret "Kill List." Several journalists involved in covering foreign affairs, including U.S. citizens, have already found themselves so listed, without any notice or warning and without any opportunity to question or challenge their inclusion. Bidal Abdul Kareem, a U.S. citizen, and Ahmad Muaffaq Zaidan, have sued the Trump administration over their listing. Both found out about their inclusion from sources other than the U.S. government.[40]

Kareem has survived at least five attempts on his life.

The speed with which present-day Americans have thrown away the hard-won protections established by their parents,' grandparents' and earlier generations, has been breathtaking.

In Canada, such a reckless, wholesale jettisoning of previous human rights safeguards hasn't yet happened, but the potential is there, thanks in part to Canada's federalist style of government, which gives its provinces considerable powers.

Historically, Canada's constitutional status has been more complex and taken longer to establish than that of its neighbour to the south. As a former—and unlike the U.S., non-rebellious—colony of England, it was established as the Dominion of Canada by the British North America Act of 1867. A series of national and international legal moves and political arguments followed, sometimes involving other British colonies and sometimes focused on Canadian federal/provincial controversies.

Then finally, in 1982, the British and Canadian governments agreed to "bring Canada's constitution home," and make Canada formally independent, via the British Canada Act and Constitution Act. The latter included the Canadian Charter of Rights and Freedoms, Canada's equivalent of the U.S. Bill of Rights.

But the Charter contained a potential spoiler, in its Section 33, the so-called "Notwithstanding Clause," inserted as a sop to provincial governments fearful that the new constitution might hamper their own lawmaking abilities. Under it, provincial legislatures or the federal government have the right to temporarily override the Charter in certain areas for a limited period of five years.

Except in Quebec, where the override was used mostly to protect French language rights, the clause has rarely been a factor.[41] It had never been invoked by the federal government or in Ontario since it was passed ... until 2018, when Conservative Ontario Premier Doug Ford invoked it. He did so to defy an Ontario Superior Court ruling concerning the size of Toronto's city council, which he wanted to shrink by executive fiat, without consulting either the council members or Toronto voters.

Ford, mockingly dubbed "Trump North" by his critics, didn't need to invoke the clause, as the Superior Court ruling that blocked him was reversed on appeal. But he later insisted "that he 'will not shy away' from invoking it again."[42] Critics pounced on Ford's move, charging it was done "in pursuit of old political grudges"[43] against certain city council members. But the damage was done, and a precedent set. How far Ford, or other politicians may go in future, using the clause to restrict basic rights, is anyone's guess.

Someday, it could be used to curb the right to dissent.

Libel and slander

Yet another brake on dissent, in this case a necessary one, are libel and slander laws, and laws against invasion of privacy. Every journalist, and every student of journalism, learns about these early on, and only the most reckless ignore them.

Webster's bare-bones definition of libel is "defamation by written or printed words, pictures or the like."[44] Basic journalism textbooks get a bit more detailed. Carole Rich, for example, quotes Donald Gillmor: "Libel is essentially a false and defamatory attack in written form on a person's reputation or character. Broadcast defamation is libel because there is usually a written script. Oral or spoken defamation is slander."[45] And the word "script" can include "headlines, photos, cartoons, film, tape, records, signs, bumper stickers and advertisements."

The key word here, of course, is "false." The basic defense against any libel or slander suit is truth. Horrid, revolting things may be published about a public figure, provided they are true and relevant. If they are, the defendant in a libel suit has a good chance of being found innocent. This author should know, having been sued for libel several times, but never convicted, because what I wrote was true, and could be proven true based on the familiar (to journalists) "rule of two sources."

This rule warns news people, before publishing or broadcasting any potentially defamatory statement, to base it whenever possible on at least two separate sources, preferably one written and one oral (from a person willing to appear in court to defend his or her comments). The written source ought to come from a respected or generally acceptable publication, and the oral source from a person of reasonably good character. The need for an oral source is not generally required when quoting already-published studies, such as textbooks or the like, whose authors and publishers have presumably already vetted what they published. Most book publishers routinely have their texts vetted by on-staff or regularly retained libel lawyers.

The published statements must also be relevant to some subject at issue. For example, noting in an article about police brutality that the local sheriff "has gooey haemorrhoids" would not be germane and would be deemed a needless invasion of privacy, unless of course it had something to do with the sheriff's brutality—as in him beating someone who poked ridicule at the state of his behind.

In the U.S., there must also be no "actual malice" involved in the publication. The U.S. Supreme Court has ruled that malice "does not mean intent to harm someone; it means that you published something knowing it was false or not bothering to check its truth or falsity."[46] The burden of proof here, fortunately, is "on the plaintiff."

The principles just stated are very general. As is the case with most laws, those on libel and slander have many variants. In the U.S., every state has its own libel laws, and in Canada various provinces have their own laws. It's always best to check first, preferably with a libel lawyer in your jurisdiction, before putting anything risky out there with your name on it. *This includes anything you may put online*. "If the defamatory statements are published—whether online or in print—they still can be considered libellous."[47]

Also, there is the matter of so-called "SLAPP suits," that is, suits intended to shut down or prohibit public discussion of an issue. SLAPP (Strategic Lawsuit Against Public Participation) suits are "intended to censor, intimidate and silence critics by burdening them with the cost of a legal defense until they abandon their criticism or opposition."[48] The *Wikipedia* entry adds:

"In the typical SLAPP, the plaintiff does not normally expect to win the lawsuit. The plaintiff's goals are accomplished if the defendant succumbs to fear, intimidation, mounting legal costs, or simple exhaustion and abandons the criticism. In some cases, repeated frivolous litigation against a defendant may raise the cost of directors and officers liability insurance ... a suit may also intimidate others from participating in the debate."

Many jurisdictions have passed legislation specifically barring such suits. Ontario, Canada, for example, passed a Protection of Public Participation Act in 2015. It noted that "A judge will dismiss the plaintiff's claim [if it] arises from an expression made by the defendant that relates to a matter of public interest."[49] This includes even informal discussions between friends or neighbours.

The idea of "fair comment" involves similar situations. "To qualify as fair comment, a statement must generally be on a matter of public interest, must be based on facts known or believed to be true, and it may not be malicious or made with reckless disregard for the truth. In this case also, truth is considered a good defense."[50] This protects, for example, critics or columnists who write negative reviews of a book, play, concert or whatnot, as well as newspaper editorials.

And if a corporation or its directors, through their business decisions, are deliberately causing severe public damage, including fatal damage to people (as with the tobacco companies), it is certainly fair comment for members of the public to publicly criticize them!

Finally, as for privacy rights, a U.S. Supreme Court decision in March 2011 ruled that corporations, despite being legal "persons," do NOT have a right to "personal privacy."[51]

~

Overall, whether in the U.S., Canada or elsewhere, there are a plethora of potential legal complications and other consequences for any public act, including dissent, and more may be coming in future government clampdowns. If you would dissent, you may wish to tread lightly.

Or, if not lightly, then very, very bravely, in defense of life itself—as those who came before us often had to do.

How they fought

Public, street demonstrations and mass marches are usually the first actions taken by people or groups with a grievance. And, particularly if they attract media attention, they may serve as an initial step to dramatize an issue and alert the public to it. Films or photos of large crowds, often including veterans in uniform wearing their medals, religious personnel wearing habits or vestments, and vehicles—EMS trucks with flashing lights, tractors operated by farmers, or the like—can make an even deeper impression.

But marchers carrying placards are all too often not enough—even the large numbers mustered by the original Occupy Wall Street demonstrators. Once the headlines are gone, they are very easily forgotten by a public inured to "splash and dash" scenes. Something more lasting is needed. Organized labour learned that long ago. Union picketers may march in the street in front of their plant, or demonstrate at key locations like railroad yards where a company ships its products, but they rarely limit their efforts to that single tactic, especially if a strike begins to drag on.

One of the first labour playbooks, and most varied set of experience-tested, and proven-successful, nonviolent direct action tactics, was probably that of the original Industrial Workers of the World (IWW) union, under its pioneering president "Big Bill" Haywood.

Founded in 1905 in Chicago, the union is still active, with branches in several countries, and has a particular interest in the environment and the issue of climate change. Characterized in its early days by both its militancy and its insistence on organizing all working people, rather than only those in certain industries or trades, it honed its skills in a series of historically significant labour conflicts.

The emergency we face today, of course, is far greater than the working conditions or salaries of the employees in a given industry. Not only are all of the employees of the major corporations, and all of their families, affected, but so are all of those companies' customers and clients, all of their families, and in fact all humans, animals and other living things. As such, everyone should have at least the moral, if not the actual legal right to employ direct action against these firms.

The IWW's online website is at https://iww.org, where it lists what the union believes are the most "Effective strikes and economic actions."[52] Many of these were first documented in writing by firebrand IWW organizer and publicist Elizabeth Gurley Flynn, who was a blunt realist when it came to bargaining with the "Robber Barons"[53] of her day. In an essay titled "Sabotage: the conscious withdrawal of the workers' industrial efficiency,"[54] she wrote:

"Labour realizes, as it becomes more intelligent, that it must have power in order to accomplish anything; that neither appeals for sympathy, nor abstract rights will make for better conditions ... It is neither sympathy nor justice that makes an appeal to the employer. But it is power. If a committee can go to the employer with this ultimatum: We [the workers] have met and formulated a demand for better hours and wages and are not going to work one day longer unless [we] get it. In other words, they have withdrawn their power as wealth producers from your plant and they are going to coerce you by their withdrawal of this power, into granting their demands."[55]

She knew, of course, that traditional strikes and simple walkouts alone might not win the battle. Staying out long enough to achieve their goals might be beyond the financial means of strikers and the strike fund of their union. She believed strikes must be well-planned, and supplemented with actions loosely described as sabotage, where the workers figuratively put on their *sabots*, or wooden shoes, mentioned earlier in this chapter, and become physically and mentally "clumsy."

"Sabotage is to the class struggle what guerrilla warfare is to the battle. The strike is the open battle of the class struggle, sabotage is the guerrilla warfare, the day-by-day warfare between two opposing classes," she wrote. Specifically what manner of sabotage, and how far to carry it was a question widely, and heatedly debated by early unionists, including Flynn.

In her view, "Sabotage means either to slacken up and interfere with the quantity, or to botch in your skill and interfere with the quality, of capitalist production or to give poor service. **Sabotage is not physical violence** [not, for instance, like the Luddites' wholesale smashing of textile mills], sabotage is an internal, industrial process. It is something that is fought out within the four walls of the shop. And these three forms of sabotage to affect the quality, the quantity and the service are aimed at affecting the profit of the employer."

That profit, she added, was the only thing that counted to the employer, whom she intuitively realized was addicted to money. "Everything is centered in his pocket book, and if you strike that you are striking at the most vulnerable point in his entire moral and economic system."[56]

To succeed, worker/saboteurs must above all be creative. "Sabotage is as broad and changing as industry, as flexible as the imagination and passions of humanity. Every day, working men and women are discovering new forms of sabotage, and the stronger their rebellious imagination is the more sabotage they are going to invent."[57]

Today, as well as historically, the IWW's toolbox reflects this view, and includes not only strikes and walkouts, but such forms of sabotage as "slowdowns, working to rule, whistle blowing (the open mouth), selective strikes, sick-ins, 'good work' strikes, sitdown strikes."[58] Some examples from the IWW website:

Slowdowns: "At the turn of the century, a gang of section men working on a railroad in Indiana were notified of a cut in their wages. The workers immediately took their shovels to the blacksmith shop and cut two inches from the scoops. Returning to work they told the boss 'short pay, short shovels.' Or imagine this: BART (Bay Area Transit authority) train operators are allowed to ask for ten 501s (bathroom breaks) anywhere along the mainline, and Central Control cannot deny them. In reality, this rarely happens. But what would management do if suddenly every train operator began taking extended 10-501s on each trip they made across the Bay?"

Working to Rule: "What would happen if ... rules and regulations were followed to the letter? French railroad strikes were forbidden [but] one French law requires the engineer to assure the safety of any bridge over which the train must pass. If after a personal examination he is still doubtful, then he must consult other members of the train crew. Of course, every bridge was so inspected, every crew member was so consulted, and none of the trains ran on time."

Whistle Blowing: "Sometimes simply telling people the truth about what goes on at work can put a lot of pressure on the boss. Consumer industries like restaurants and packing plants are the most vulnerable ... Waiters can tell their restaurant clients about various shortcuts and substitutions that go into creating the faux-haute cuisine being served to them."

Employees of some of the most destructive corporations have often leaked internal documents or confidential information, such as the 1980s Exxon study mentioned in Chapter 4, that warned of "catastrophic" climate events, leaked in 2015.

Selective Strikes: "Unpredictability is a great weapon in the hands of the workers. Pennsylvania teachers used the Selective Strike in 1981, when they walked a picket line on Monday and Tuesday, reported for work on Wednesday, struck again on Thursday, and reported for work on Friday and Monday. This not only prevented administrators from hiring scabs to replace the teachers, but also forced administrators who hadn't been in a classroom for years to staff the schools while the teachers were out."

Though not mentioned specifically on the IWW site, selective and rotating strikes, focusing on first one, then another firm in a given industry, have been employed in many labour disputes to put pressure on a single company, while allowing its competitors in the same industry to function normally. Pouncing on the opportunity, competitors may poach the first company's clients, and retain them after the strike is over. That prospect often makes the first company more amenable to the strikers' view.

Sick-Ins: "A good way to strike without striking. The idea is to cripple your workplace by having all or most of the workers call in sick on the same day or days. Unlike the formal walkout, it can be used effectively by single departments and work areas ... It is the traditional method of direct action for public employee unions, which are legally prevented from striking."

'Good Work' Strikes: "Many forms of direct action, such as slowdowns, end up hurting the consumer ... One way around this is to provide better or cheaper service—at the boss' expense, of course. Workers at Mercy Hospital in France, who were afraid that patients would go untreated if they went on strike, instead refused to file the billing slips for drugs, lab tests, treatments and therapy. As a result, the patients got better care (since time was spent caring for them instead of doing paperwork), for free. The hospital's income was cut in half, and panic-stricken administrators gave in to all of the workers' demands."

Sitdown Strikes: "IWW theater extras, facing a 50 per cent pay cut, waited for the right time to strike. The play had 150 extras dressed as Roman soldiers to carry the Queen on and off stage. When the cue for the Queen's entrance came, the extras surrounded the Queen and refused to budge until the pay was not only restored, but tripled."

And, as described earlier, the United Auto Workers were engaged in a sitdown strike, which brought Ford auto production lines to a halt, when the legendary Battle of the Overpass took place. They, too, won their demands.

Far more complete an outline of possible nonviolent methods, even more detailed than the IWW list, was put together by Peter Ackerman and Christopher Kruegler in their 1994 book, *Strategic nonviolent conflict: the dynamics of people power in the 20th century.*[59] They even produced a table, divided into three columns under "1) Protest and persuasion, 2) Non-cooperation and 3) Intervention,"[60] offering no fewer than 42 types of actions! This book, of which more will be said below, is well worth consultation.

Other tactics were outlined in an Op-Ed article in 2014 by Nadine Bloch, "Nine extraordinary ways to use the tools of your trade in protest."[61] These included musicians pinning protest signs to their instruments and playing outdoor concerts, to posting billboards with strikers' messages or blockading roads with columns of slow-moving tractors (to protest wholesale low milk prices to farmers). Though not mentioned in the article, benefit concerts by celebrity entertainers have also been employed to gain support for causes, as the appearances of pro-labour folk singers like Pete Seeger have shown.

Public relations: Strike propaganda has taken a myriad forms, from leaflets, flyers tacked to posts, mass or neighbourhood-specific direct-mailings and door-to-door handouts, to outdoor billboards posted in strategic locations, to online polemics posted on websites or blogs. Targets of a strike, from company CEOs to hired strikebreaking scabs might sometimes be named in these, rather than naming only the corporation, as long as anything said in the texts is true, according to the "rule of two sources," and relevant to the public issue involved. But check with a lawyer beforehand.

Humour is also an effective tool, as a giant, rubber chicken with orange hair deployed near the White House to embarrass U.S. President Donald Trump demonstrated.[62] Horror can be equally effective. One group of Vegan activists played films of animals being slaughtered outside of restaurants where the dead animals' meat was being served.[63]

Where propaganda methods are concerned, a philosophical suggestion, born of this author's experience with martial arts, may also be in order. To the uninitiated observer, an Aikido practice session might look violent. In reality, it is the opposite, despite the rapid movements involved and the loud thumps when someone hits the mat. The techniques developed by Ueshiba Morihei, founder of the discipline, are essentially defensive, not aggressive, and any violence present in them originates with the aggressor, not the defender.

Summed up, essentially, as "step in, turn, disperse," Aikido techniques take an aggressor's energy and deflect or reverse it, turning it back on him, allowing him to expend his own violence harmlessly—or at least harmless to his intended victim. If the attack is violent enough, the attacker may hurt himself. But the *aikidoka* (practitioner of Aikido) didn't hurt him. He, or she, just got out of the way. Aikido means "the way of harmony," and its best techniques are those that injure no one, leaving both attacker and defender intact—and perhaps a bit wiser.

Fighting the public relations battle can use a similar approach. For example, if a company advertises a product that is destructive, *amplify* their ads, by giving the whole story, including the product's pernicious effects. Spread this, with the company's name attached, as widely as possible, just as cigarette packages now have warnings with the image of a girl with sunken eyes and rotting teeth, etc. Make labels and stick them on packages, post billboards mocking the product, publish ads in papers or online, send out flyers in mass mailings, all touting the destructive wonders of the fabulous whatever-it-is.

Company directors currently hide behind the "legal person" concept and tout products as sold by or made by the company, not them. The *Joe Camel* cartoon character of tobacco wars history took this a step further, giving the company an actual, ridiculous face. Such cartoon characters might do damage here, there and everywhere, or lead a parade of actual board members, CEOs and the leading investors/hedge funds that back them, mocking their "accomplishments" in producing products that kill people (tobacco), wipe out honeybees, destroy agriculture, or indebt students.

Billboards posted strategically near corporate headquarters, or en route to the country clubs where directors go to play golf, might tout the negative feats of various corporate officers.

The concept has been used for decades in countless newspaper editorial cartoons, lampooning politicians and other public figures. Such cartoons typically exaggerate features of their targets, taking their actions to extremes—much as the *aikidoka* exaggerates his or her opponent's violence. Just be sure to consult with a lawyer first, and limit the underlying message to what is factually true and can be backed up in court. Play it straight. All is NOT fair in love and war.

Boycotts: Mass refusal to purchase a company's products or services, and urging the public in general not to purchase them, can have a sharp and rapid effect on profits, and help bring company officers to the bargaining table. And, of course, there is nothing illegal about simply not buying something. *Urging others* not to buy, however, may be actionable in court, under the headings of harassment or restraint of trade, depending on one's jurisdiction. Again, a lawyer's advice should be sought.

Blockades: Physically preventing actions that damage the public or offend public values can take a variety of forms, the most basic of which is the simple blockade. The residents of Stroud, England, for example, were alarmed when plans for a new supermarket required widening a road and felling "a whole row of magnificent lime trees. To ward off power saws, local protesters wrapped the trees in steel mesh. They then built platforms in the branches, 30 feet above the ground. The trees were guarded around the clock, and whenever 'officials' appeared, the campaigners climbed up to their platforms and pulled the ladders up behind them. The stunt worked, and after weeks of delay, the decision was made to spare the trees and find an alternative way of managing the traffic."[64]

In another action, Greenpeace activists, demanding an end to all new oil and gas exploration, brought heavy containers, each weighing several tonnes, to blockade "all five entrances to BP headquarters in London" just before the company's annual general meeting in May 2019. Admitting the tactic was illegal, one protester said: "It's highly likely we'll be arrested eventually," but added: "We're shutting down BP's HQ because business as usual is just not an option. BP is fuelling a climate emergency that threatens millions of lives and the future of the living world."[65]

Class action lawsuits: While in the lawyer's office asking about protesters' potential liability for damages, it would also be a good time to enquire about class-action lawsuits against the corporations, first to make them cease and desist their ongoing noxious activities, and second to make them pay damages. Though few ordinary people can easily afford today's legal fees and court costs, many individuals, pooling their resources as a group, might have deeper pockets. A series of class-action damage suits, brought against several corporate defendants, perhaps by several concerned groups, could be as effective as a strike, and even more time-consuming for any defendants. Crowd-funding, which has become an option especially in this age of online communication, is another possible source of funds.

In order to enrich themselves, corporations have knowingly and deliberately caused damage, including fatal damage, on a colossal scale, to individuals, communities and the environment. And, since many of them knew beforehand from their own internal studies exactly what the effects of their actions would be, the damage was inflicted "intentionally or recklessly." It could thus qualify as criminal damage under various statutes in many countries.

For example, the UK's 1971 Criminal Damage Act states that anyone "who, without lawful excuse, destroys or damages any property belonging to another, intending to destroy or damage any such property, or being reckless as to whether any such property would be destroyed or damaged" commits a criminal offense.[66] Where the damage is more than 5,000 pounds Sterling, "the maximum sentence is 10 years imprisonment," and "where there was also an intention to endanger life, either intentionally or recklessly, the maximum sentence is life imprisonment."

Of course, one can't imprison a corporation, though perhaps some of its directors might be vulnerable. The physical damage caused, however, may call for both monetary "compensation" and/or "restitution." In common legal terminology, compensation is paid to cover the victim's actual losses, while restitution is the payment of any gains realized by the criminal from his, her, their or its destructive actions.[67] The latter usually comes under the heading of "unjust enrichment," or profiting from an unlawful act.

"Common law systems such as those of England, Australia, Canada and the United States typically adopt the 'unjust factor' approach. In this analysis, the claimant must point to a positive reason why the defendant's enrichment is unjust."[68] One would think that making money by deliberately killing innocent people, ruining the soil, or by destroying the entire planet's environment, making it unliveable, ought logically to qualify.

Ackerman and Kruegler, mentioned above, emphasize this in their study, which examined several nonviolent real-world campaigns, including the Danish resistance to the Nazis in the Second World War, Solidarity's initial, 1980-81 battle against the Polish Communist Party, the First Russian Revolution against the Tsars, in 1904-06, and other historical contests featuring nonviolent campaigns.

Their Russian example, in which a loose coalition of urban industrial workers, peasants, students and others first attempted to alter unjust economic and social conditions by naively trying to persuade their "little father" the Tsar to listen to their pleas, shows what can happen when clear agreement on goals and a long-range strategic plan are lacking.

The rebellion began with a peaceful march by up to 150,000 people, heading for the ruler's Winter Palace. They were met by mounted cavalry, which first attacked using whips, clubs and the flats of swords, then fired on the marchers, killing at least 200 and wounding 800 or so more.[73] Thus graphically disabused of their little father's kindliness, the rebels regrouped and launched a longer campaign which, though it weakened the Tsar's legitimacy and paved the way for a later, violent revolt and the advent of the Bolsheviks, ended in short-term failure.

On policy, some rebels wanted to replace the Tsar altogether, while others wanted only to persuade him to make reforms. Others favoured a strategy aimed at democratizing the country, using the *Duma* or legislature as the means, while others saw electoral politics as a waste of energy. As the authors summarize:

> "The opposition movement, as a whole, had no grand design to guide its activities. It lacked both a capacity for comprehensive planning and a solid consensus on ultimate political objectives. At certain points, however, the initiative of a single faction of the movement became decisive in determining the future course of the whole. Sometimes this would occur on the basis of explicit strategic reasoning, but more often it was a case of reacting to a particular crisis or challenge in a way that produced unexpected results."[74]

The authors list five *Principles of Development*, four *Principles of Engagement* and three *Principles of Conception*, and find the early Russian rebels deficient in most of them.[75] Unfortunately for Russia and the world, their communist Bolshevik successors, under Lenin, Trotsky and others, were not deficient. The tyranny of Stalin and long communist nightmare followed. Details of this analysis are too lengthy to go into here, but anyone seriously hoping to change our fragile present world for the better would be well advised to look them up, and heed them.

A more recent book-length look at tactics has been published independently online by Roger Hallam, titled: *How to win! Successful procedures and mechanisms for radical campaign groups: Radical Think Tank*. A short, but sensible look at the subject was also published by Olga Khazan in *The Atlantic*, "The psychology of effective protest."[76]

Geo-engineering

Finally, as the fight to slow, or even reverse, the ongoing ruin of our planet continues, a number of suggestions have been, and will continue to be made regarding ways to physically reverse some of the most destructive processes. Geo-engineering, the discipline that groups them together, could prove helpful.

For example, scientists recently discovered that a mineral, hydrotalcite, often treated as waste material at mine sites, can react with carbon dioxide, trapping the gas in crystal form. Their study, published in 2018, looked at tailings from an old asbestos mine in Australia.[77] "As rainwater eats away at some of this mine waste, magnesium reacts with the CO_2 in the air to form hydrotalcite, capable of storing the greenhouse gas." Large amounts of the mineral might be used in strategic locations to trap the gas most responsible for climate change, making it harmless.

There are many other possibilities, which readers of the *International Journal of Greenhouse Gas Control*[78] read about in each issue. It is well worth a subscription.

Before lumping all our eggs in any "quick-fix," geo-engineering basket, however, stop and reflect: Who is offering the latest panacea? Are the corporations its source? Will they profit from it? Chapter One

19 *Op. Cit.*, 2.

20 Hilf & Hilf, PLC, "Malicious destruction of property," *Justicia blog*, as posted online 25 December 2018 at www.michigancriminalattorneysblog.com/malicious-destruc ...

21 *Loc. Cit.*

22 *Wikipedia, the Free Encyclopedia*, "Thoughtcrime," as posted online 4 December 2018 at https://en.wikipedia.org/wiki/Thoughtcrime, 1-2.

23 *United States Constitution*, First Amendment: "Congress shall make no law ... prohibiting .... the right of the people peaceably to assemble, and to petition the government for a redress of grievances," as posted online 5 December 2018 at www.usconstitution.net/xconst_Am!.html

24 Emile Pouget, *Le sabotage*, (FB Editions, Middletown, Delaware, 2017), reprint of a text first published in 1911.

25 Jenny Nolan, "The battle of the overpass," 7 August 1997, *Detroit News*, as cited in Wikipedia, the Free Encyclopedia, "Battle of the Overpass," and posted online 3 December 2018 at https://en.wikipedia.org/wiki/Battle_of_the_Overpass.

26 Two of the best works on Jean Moulin and the *Maquis* in general are: Jean-Pierre Azema, *Jean Moulin*, (Paris, Editions Perrin, 2006); and Daniel Cordier, *Jean Moulin: la republique des catacombes*, (Paris, Editions Gallimard, 1999).

27 *Random House Webster's College Dictionary*, (New York, Random House, 1995), 1161.

28 *The Oxford English Reference Dictionary*, (Oxford, Oxford University Press, 1996), 1242.

29 *The Free Dictionary, Legal Dictionary*, "Riot, legal definition of riot," as posted online 19 October 2018 at https://legal-dictionary.thefreedictionary.com/riot,1.

30 *Wikipedia, the Free Encyclopedia*, "Riot," as posted online 19 October 2018 at https://en.wikipedia.org/wiki/Riot, 4.

31 M. Cherif Bassiouni, *The law of dissent and riots*, (Springfield, Illinois: Charles C. Thomas, Publisher, 1969).

32 Bassiouni, *Op.Cit.*, 497.

33 *Op. Cit.*, 498.

34 *Op. Cit., Wikipedia,* "Riot," 3.

35 Michael Ratner and Margaret Ratner Kunstler, *Hell no: your right to dissent in twenty-first-century America*, (New York: The New Press, 2011), 87-88.

36 *Op. Cit.*, 88.

37 *Ibid.*, 88-89.

38 *Op. Cit.*, 45.

39 *Op. Cit.*, 17.

40 Marjorie Cohn, "How do you get off the U.S. 'Kill List'?" 29 May 2018, *Truthout*, as posted online 30 May 2018 at www.truth-out.org/news/item/44620-how-do-you-get-off-the-us- ..., 1.

41 Richard Poplak, "A populist has exposed a sinkhole in Canada's democracy," 28 September 2018, *The Atlantic*, as posted online 19 December 2018 at www.theatlantic.com/international/archive/2018/09/a-populi ..., 4-5.

42 *Op. Cit.*, 5.

43 Andrew Russell, "Ontario Premier Doug Ford plans to invoke notwithstanding clause. Here's what you need to know," 10 September 2018, *Global News*, as posted online 18 December 2018 at https://globalnews.ca/news/4438198/notwithstanding-clause-doug- ... 3.

44 *Random House Webster's College Dictionary*, (New York, Random House, 1995), 780.

45 Carole Rich, *Writing and reporting news*, (Belmont, California, Wadsworth/Thomson Learning, 2003), 306-7.

46 *Op. Cit.*, Rich, 308.
47 *Op. Cit.*, Rich, 307.
48 *Wikipedia, the Free Encyclopedia*, "Strategic lawsuit against public participation," as posted online 1 December 2018 at https://en.wikipedia.org/wiki/Strategic_lawsuit_against_public_part ..., 1.
49 *Canadian Civil Liberties Association*, "Ontario legislation: the Protection of Public Participation Act, 2015," as posted online 7 October 29017 at http://rightswatch.ca/2015/11/10/ontario-legislatuion-the-protection-of ..., 2.
50 *Op. Cit.*, Rich, 313.
51 *Oyez.org*, "FCCv. AT&T Inc.," March2011, as posted online 21 December 2018 at ://www.oyez.org/cases/2010/09-1279
52 *Industrial Workers of the World*, "Effective strikes and economic actions," as posted online 12 December 2018 at https://iww.org/about/solidarityunionism/directaction
53 *Wikipedia, the Free Encyclopedia*, "Robber baron (industrialist)," as posted online 25 December 2018 at https://en.wikipedia.org/wiki/Robber_baron_(industrialist).
54 Elizabeth Gurley Flynn, Walker C. Smith, William E. Trautmann, *Direct action and sabotage: three classic IWW pamphlets from the 1910s*, (Oakland, California, Charles H. Kerr Library/PM Press, 2014), 92-3.
55 *Loc. Cit.*
56 *Op. Cit.*, 94.
57 *Op. Cit.*, 113.
58 Industrial Workers of the World, *Op. Cit.*, "Effective strikes," 1.
59 Peter Ackerman and Christopher Kruegler, *Strategic nonviolent conflict: the dynamics of people power in the 20th century*, (Westport, Connecticut, Praeger Publishers, 1994).
60 *Op. Cit.*, 8.
61 Nadine Bloch, "Nine extraordinary ways to use the tools of your trade in protest," 31 August 2014, *Truthout*, as posted online 31 August 2014 at http://truth-out.org/opinion/item/25913-nine-extraordinary- ways-to-use ...
62 Nicholas Hautman, "Giant, inflatable chicken with Donald Trump-like hair pops up near White House," 9 August 2017, *US Magazine*, as posted online 30 December 2018 at www.usmagazine.com/celebrity-news/inflatable-chick ...
63 Debayan Paul, "Vegan activists campaigned at Nando's; played slaughtering footages," 29 May 2019, *Raise Vegan*, as posted online 29 June 2019 at https://raisevegan.com/vegan-activists-campaigned-at-nandos-play ..., 2.
64 Chris Baines, *101 ways to really save the world*, a *BBC Wildlife Magazine* Ultimate Guide, March 1993, *BBC Wildlife Magazine*, Bristol, UK, 23.
65 Matthew Weaver, "BP headquarters in London blockaded by Greenpeace," 20 May2019, *The Guardian*, as posted online 20 May 2019 at www.theguardian.com/business/2019/may/20/bp-headquarte ..., 1.
66 Seatons Law Ltd., "Criminal Damage," as posted online 11 July 2019 at www.seatons.co.uk/legal-services/criminal-law/criminal-damage-offences/, 1.
67 *Wikipedia, the Free Encyclopedia*, "Restitution," at posted online 5 July 2019 at https://en.wikipedia.org/wiki/Restitution, 1.
68 *Wikipedia, the Free Encyclopedia*, "Unjust enrichment," as posted online 5July2019 at https://en.wikipedia.org/wiki/Unjust_enrichment, 1.
69 See: Parmy Olson, *We are Anonymous*, (London, William Heinemann, 2012).
70 See: Luke Harding, *The Snowden files: the inside story of the world's most wanted man*, (London, faber & Faber, 2014)
71 *Wikipedia, the Free Encyclopedia*, "Internet activism," as posted online 29 June 2019 at https://en.wikipedia.org/wiki/Internet_activism

Masanobu Fukuoka, *The Natural Way of Farming: the theory and practice of Green philosophy*, (Tokyo, Japan Publications Inc., 1985).
Written by Japan's internationally famed farmer-philosopher, this is an introduction to his understanding of organic farming and its place in the world. His vision goes far beyond most organic farming how-to's, looking deeply into the reasons for each practice.

Masanobu Fukuoka, *The Road Back to Nature: regaining the paradise lost*, (Tokyo, Japan Publications Inc., 1987).
The sequel to the *Natural Way*, this book is as much about human nature as anything else.

Donald Worster, *Dust Bowl: the southern plains in the 1930s*, (Oxford, Oxford University Press, 1979).
This is just how bad it can get, by an author whose parents lived through the worst of the Dirty Thirties' Dustbowl.

Judith Shapiro, *Mao's War Against Nature: politics and the environment in revolutionary China*, (Cambridge, Cambridge University Press, 2001).
How NOT to do it, in farming and everything else, as demonstrated by the ideological fanatics working under Mao Tse Tung's communists during the so-called "Great Leap Forward" years. A lesson in human extremes.

THE ECONOMIC SYSTEM

Books

Thomas F. Pawlick, *Debt Sentence: how Canada's student loan system is failing young people and the country*, (Bradenton, Florida, BookLocker.com, Inc., 2012).
Alan Michael Collinge, *The Student Loan Scam: the most oppressive debt in U.S. history—and how we can fight back*, (Boston, Beacon Press, 2009).
Two dissections of the student loan racket currently impoverishing young people around the world, one focused on Canada, the other on the U.S.

James D. Scurlock, *Maxed Out: hard times, easy credit and the era of predatory lenders*, (New York, Scribner, 2007).

A journalistic expose of the consumer loan industry in the U.S., with special attention to the outrageous and the predatory.

E.E. Schumacher, *Small is Beautiful: a study of economics as if people mattered*, (London, Blond & Briggs Ltd., 1973).

Donella H. Meadows, Dennis L. Meadows, Jorgen Randers, William W. Behrens III, *The Limits to Growth: a report for the Club of Rome's project on the predicament of mankind*, (New York, New American Library, 1972).

Two classics of the 1960s-70s "movement years," these books set the tone for debate in the first serious, sustained questioning of the organization of society and the world economy since the Great Depression.

Noam Chomsky, *The Prosperous Few and the Restless Many*, (Berkeley, California, Odonian Press, 1993).

A hard look at who really benefits from NAFTA, and the international economy in general, as well as a good introduction to the writing of Chomsky, a.k.a. Everybody's Intellectual.

Morton Mintz and Jerry S. Cohen, *America, Inc.: who owns and operates the United States*, (New York, Dell Publishing Co., Inc., 1972).

As the Kansas City *Times* reviewer described it: "corporate efforts to win protected positions, regulate the regulators, squeeze out the small operators, take over the markets, administer prices, buy political influence."

Howard Zinn, *A People's History of the United States: 1492–present*, (New York, HarperCollins Publishers, 2001).

Iconoclastic historian Zinn writes the saga of the USA from the standpoint of those at the bottom, rather than those on top.

David C. Korten, *When Corporations Rule the World*, (San Francisco, California and West Hartford, Connecticut, Kumerian Press Inc. And Berrett-Koehler Publishers, 1995).

An expose of corporate economic globalization, written by a well-credentialed critic.

Susan Jacoby, *The Age of American unreason*, (New York, Pantheon Books, 2008).

An in-depth look at the "dumbing down" of American life that has paved the way for so much political and social craziness, from the "denialist" phenomenon and rejection of science, to the election of nincompoops. Her picture of North Americans as they were in earlier generations, when even ordinary, working-class folks respected learning and spent a considerable time educating themselves, is one this author recalls with nostalgia.

Naomi Oreskes & Erik M. Conway, *Merchants of Doubt: how a handful of scientists obscured the truth on issues from tobacco smoke to global warming*, (New York, Bloomsbury, 2010).

An expose of the denial industry, from its birth during the "tobacco wars" to today, when all science is routinely denied and ridiculed, and reading any book is seen as a subversive activity.

Jacques Ellul, *Propaganda: the formation of men's attitudes*, (New York, Vintage Books, 1965).

How and why lies and falsehoods have misled and continue to mislead individuals and public opinion.

Videos

Prof. Jeffrey L. Kasser, *Philosophy of Science*, (Chantilly, Virginia, The Teaching Company, 2006).

One of a series of video productions titled *The Great Courses*, offered by The Teaching Company, this university-level series of lectures provides an in-depth look at the reasoning process that gave rise to modern science and its place in society.

Prof. Richard Wolfson, *Earth's Changing Climate*, (Chantilly, Virginia, The Teaching Company, 2007).

One of a series of video productions titled *The Great Courses*, offered by The Teaching Company, this course outlines the structure and long-term history of our planet's climate system, including the recent effects of climate change.

Daniel B. Gold and Judith Helfand, directors, *Everything's Cool: a toxic comedy about global warming*, (St. Laurent, Quebec, City Lights Pictures, 2007).

Black comedy, based (loosely) on what really might happen.

MULTI-SYSTEM COLLAPSE

Books

The Holy Bible, Book of Revelation (a.k.a. the Apocalypse): any version you prefer.

Video

Prof. Craig R. Koester, *The Apocalypse: controversies and meaning in western history*, (Chantilly, Virginia, *The Teaching Company*, 2011).

One of a series of video productions titled *The Great Courses*, offered by The Teaching Company, this course examines the various interpretations given to the Book of Revelation, emphasizing those of impartial, present-day scholars.

THOSE RESPONSIBLE

Books

Norman Ohler, *Blitzed, Drugs in Nazi Germany*, (London, Penguin Books, 2017).

A look at the last time a group of addicts, with a fascist worldview, threatened (nearly) the entire planet.

FIGHT THE GOOD FIGHT

Books

Marcus Tullius Cicero, *On Obligations* (*De officiis*), (Oxford, Oxford University Press, 2000).

The famed Roman senator's views on civic duty, up to and including war, written as advice to his son, Marcus.

Henry David Thoreau, *Walden, or a life in the woods, and the famous essay on Civil Disobedience,* (New York, New American Library, 1960).
Originally published in 1849, Thoreau's essay "On the duty of civil disobedience," included in this collection, inspired future generations. The essay, as well as Thoreau's experiences at Walden, influenced both Gandhi and Martin Luther King, Jr.

Mohandas K. Gandhi, *An autobiography: the story of my experiments with truth,* (Boston, Beacon Press, 1993).
The leader of India's successful, nonviolent battle for independence from the British, tells his life story, and how his thinking led to the struggle that changed his nation's history.

Clayborne Carson, editor, *The Autobiography of Martin Luther King, Jr.*, (New York, Grand Central Publishing, 2001).
The U.S. civil rights crusader's papers and writings, collected and edited in one volume.

Stephane Hessel, *Indignez-vous!* (Montpellier, France, Indigene Editions, 2011).
A former member of the French Resistance against Nazi occupation in the 1940s urges today's youth to rise up and fight for their rights, just as the original *Maquis* fought for theirs.

Jean-Pierre Azema, *Jean Moulin: le politique, le rebelle, le resistant,* (Paris, Editions Perrin, 2003).
The story of a leader of the French Resistance, De Gaulle's representative on the scene who gave his life for freedom.

Erica Chenoweth and Maria J. Stephan, *Why Civil Resistance works: the strategic logic of nonviolent conflict,* (New York, Columbia University Press, 2011).
A scholarly, yet very readable look at nonviolent political resistance movements around the world, and why so many succeed.

Peter Ackerman and Christopher Kruegler, *Strategic Nonviolent Conflict: the dynamics of people power in the 20th century*, (Westport, Connecticut, Praeger Publishers, 1994).

An examination of several well-known nonviolent resistance movements and why so many did NOT succeed.

Elizabeth Gurley Flynn, Walker C. Smith and William E. Trautmann, *Direct Action and Sabotage: three classic IWW pamphlets from the 1910s*, (Oakland, California, PM Press, 2014).

The tactics and thinking behind the early labour struggles of the Industrial Workers of the World (IWW), who set the pattern for most militant union agitating since.

Emile Pouget, *Le Sabotage*, (Middletown, Deleware, FB Editions, 2017).

Published originally in 1911, this was a history of labour sabotage as practiced in France from 1897 to 1911, including how it got its name.

Michael Ratner and Margaret Ratner Kunstler, *Hell NO: your right to dissent in 21st century America*, (New York, The New Press, 2011).

An excellent resume of how human rights, particularly the right to dissent, have been gradually eroded in the U.S. since the terrorist attacks of 9-11, ushering in an era of near-lawlessness. Advice to would-be protesters on how to deal with such matters as violent repression, arrest and imprisonment is included.

M. Cherif Bassiouni, *The Law of Dissent and Riots*, (Springfield, Illinois, Charles C. Thomas, Publisher, 1971).

Now fairly dated, this remains nevertheless one of the most complete outlines of the legal details of "riot" and public disturbances, giving definitions of the terms involved, and the various punishments on the books in the U.S. and its individual states.

Poh-Gek Forkert, *Fighting Dirty: how a small community took on big trash*, (Toronto, Between the Lines, 2017).
The impassioned tale of how the ordinary residents of a tiny rural area in eastern Ontario, Canada, took on a major polluter and won, thanks to good organization, innovation and especially community spirit.

John Greenwood, *Worker sit-ins and job protection*, (Westmead, Hants, Gower Press, 1977).
Case studies from Europe, especially France, on union action to prevent factory closures and job loss.

Parmy Olson, *We are Anonymous: inside the hacker world of LulzSec, Anonymous and the global cyber insurgency*, (London, William Heinemann, 2012).
A voluminous, but well-written discussion of the new frontier of online hacking and rebellion.

Luke Harding, *The Snowden files: the inside story of the world's most wanted man*, (London, Faber & Faber, 2014)
A recounting of Edward Snowden's release of top secret files from the U.S. National Security Agency (NSA), its worldwide impact and the price he has paid personally since.

Kisshomaru Ueshiba, *The spirit of Aikido*, (Tokyo, Kodansha International, 1984).
Written by the son of Ueshiba Morihei, founder of Aikido, this book describes the history and philosophy behind what many consider the greatest of the Japanese martial arts.

U.S. Army, *The Official U.S. Army Small Unit Tactics Handbook: infantry platoon and squad*, (Fort Benning, Georgia, Carlile Media, 2018).
If the U.S. government decides to use troops, or perhaps militarized police, to put down citizen protests violently, here are the tactics they may use.

Murray Bookchin, *Remaking Society: pathways to a green future*, (Boston: South End Press, 1990).

An argument for "an ecological society based on nonhierarchical relationships, decentralized, democratic communities, and eco-technologies like solar power, organic agriculture and humanly-scaled industries."

Periodical

Sean McCoy, editor-in-chief, *International Journal of Greenhouse Gas Control*, (Elsevier Publications).

A peer-reviewed scientific journal devoted to finding ways to solve the climate change problem. Its website is at www.journals.elsevier.com/international-journal-of-greenho ...

Video

Richard Attenborough, director, *Gandhi: his triumph changed the world forever*, (Culver City, California, Columbia Pictures, 1982).

The biography of Mohandas K. Gandhi and his nonviolent liberation of India from the British. The film won eight Academy Awards.

About the Author

Thomas Pawlick's qualifications include 50 years as a science journalist, with 10 as a university professor of journalism. His work has won numerous national and international awards, including the U.S. Brotherhood Award and the Canadian Science Writers Association National Journalism Award (three times). He has published 10 previous books, one of which, *The End of Food*, was a bestseller.

Printed in May 2020
by Gauvin Press,
Gatineau, Québec